Published in Canada by Fifth House Ltd.
A Fitzhenry & Whiteside Company
195 Allstate Parkway
Markham, ON L3R 4T8
www.fifthhousepublishers.ca

Published in the United States by Fifth House Ltd.
A Fitzhenry & Whiteside Limited Company
311 Washington Street, Brighton, Massachusetts 02135

Library and Archives Canada Cataloguing in Publication
Schacter, Mark, 1957-
Roads / photographs by Mark Schacter.
ISBN 978-1-897252-60-4
1. Roads—Canada—Pictorial works. 2. Landscape photography—Canada. I. Title.
TR660.5.S33 2010 779'.3671092 C2010-904583-1

Fifth House Ltd. acknowledges with thanks the Canada Council for the Arts, and the Ontario Arts Council for their support of our publishing program. We acknowledge the financial support of the Government of Canada through the Book Publishing Industry Development Program (BPIDP) for our publishing activities.

Edited by Meaghan Craven
Design by BookWorks

Printed in Canada
10 9 8 7 6 5 4 3 2 1

ONTARIO ARTS COUNCIL
CONSEIL DES ARTS DE L'ONTARIO

Canada Council for the Arts
Conseil des Arts du Canada

ROADS

Mark Schacter

FIFTH HOUSE

Published in Canada by Fifth House Ltd.
A Fitzhenry & Whiteside Company
195 Allstate Parkway
Markham, ON L3R 4T8
www.fifthhousepublishers.ca

Published in the United States by Fifth House Ltd.
A Fitzhenry & Whiteside Limited Company
311 Washington Street, Brighton, Massachusetts 02135

Library and Archives Canada Cataloguing in Publication
Schacter, Mark, 1957-
Roads / photographs by Mark Schacter.
ISBN 978-1-897252-60-4
1. Roads—Canada—Pictorial works. 2. Landscape photography—Canada. I. Title.
TR660.5.S33 2010 779'.3671092 C2010-904583-1

Fifth House Ltd. acknowledges with thanks the Canada Council for the Arts, and the Ontario Arts Council for their support of our publishing program. We acknowledge the financial support of the Government of Canada through the Book Publishing Industry Development Program (BPIDP) for our publishing activities.

Edited by Meaghan Craven
Design by BookWorks

Printed in Canada
10 9 8 7 6 5 4 3 2 1

ONTARIO ARTS COUNCIL
CONSEIL DES ARTS DE L'ONTARIO

Canada Council for the Arts
Conseil des Arts du Canada

They stand uncertainly underneath immense skies, and everything about them is drowned. Where go? What do? What for?—sleep. But this foolish gang was bending onward.

—Jack Kerouac, *On the Road*

In matters such as the appearance of a nation, the truth itself is controversial. Photography, that machine recorder, is subjective in the hands of a man who has something to be subjective about.

—Walker Evans, photographer, 1903–1975

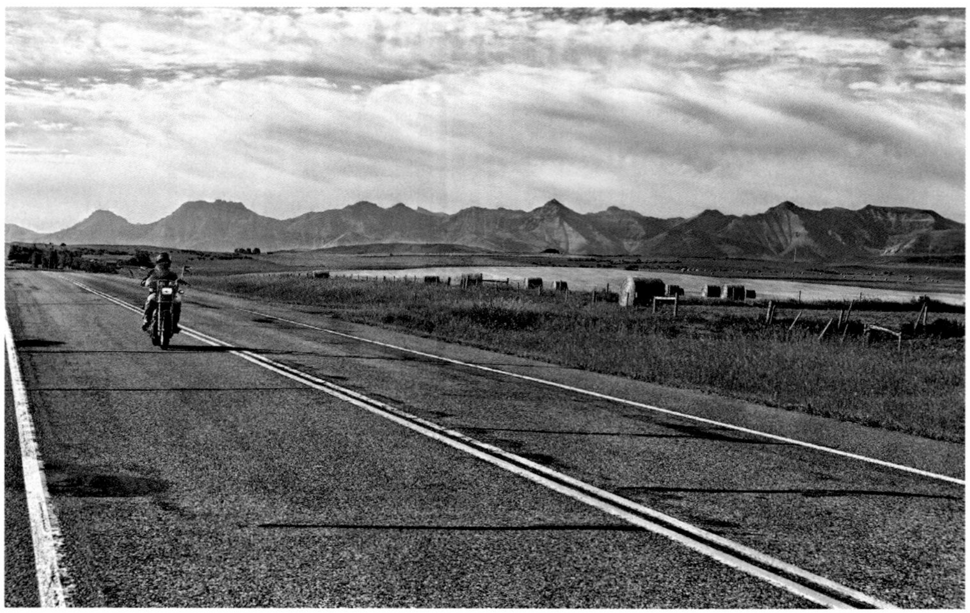

Contents

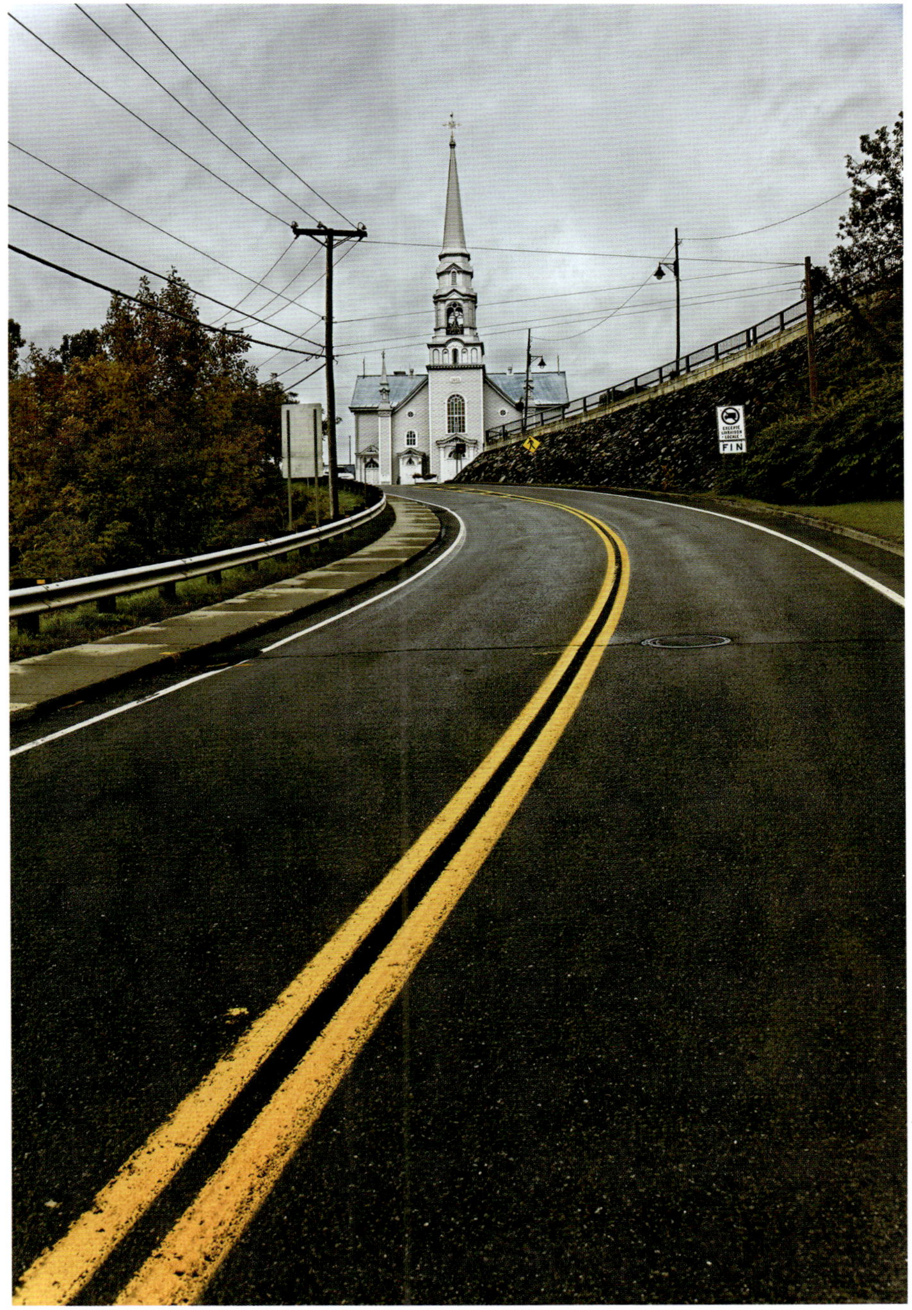

Photographer's Preface

1

WALKER EVANS, the American photographer of the mid-1900s, once said that "valid photography, like humor, seems to be too serious a matter to talk about seriously." Photography's power comes from the possibility of communicating ideas or emotions in ways that words can't, so any attempt to explain in detail what a photograph is supposed to be about or why it was taken is doomed from the start.

The problem of translating images into words is compounded when the photographs are your own. You can try to stand outside the photographs and analyze them "objectively," as if they were someone else's—an impossible task. Or you can burrow into your own head and try to explain, to yourself as much as to the reader, what you were thinking when you took the photographs—which is likely to lead to an endless argument between different parts of yourself. And surely, in any case,

the viewer needs to be left some private space for contemplation and enjoyment unencumbered by the photographer's interpretive gloss.

One of the most important photographers of the twentieth century, Robert Frank (who, as I write this, is in his eighties and lives in Mabou, Nova Scotia), has always refused to explain his work, arguing that the viewer "must have something to see. It is not all explained for him." The fact that a photograph can simultaneously appear "realistic" and mysterious is indeed what makes it exciting to take photographs and to look at them.

On the other hand, as a consumer of photographic images, I appreciate the value of direct insight into the photographer's thought process. And so it's with these conflicting ideas in mind that I have written the commentary that you will find interspersed with the photographs in *Roads*. As a result, I have tried to provide a balance of description and silence throughout.

2

Canada comprises nearly 10 million square kilometres of territory that is inhabited by just 34 million people. It is one of the emptiest countries on Earth in terms of human habitation. Of 239 countries in the world, only eleven of them are less densely populated than Canada.

I am drawn to emptiness: its look and sound and feel. And I am fascinated by signs of human effort—which may be variously poignant, heroic, magnificent, tragic, comic, ridiculous, or fruitless—to fill the void. As an archaeologist might be, I'm interested in experiencing landscapes as being full of evidence that people have been there, built something, scratched a living out of the earth, come, and gone, to

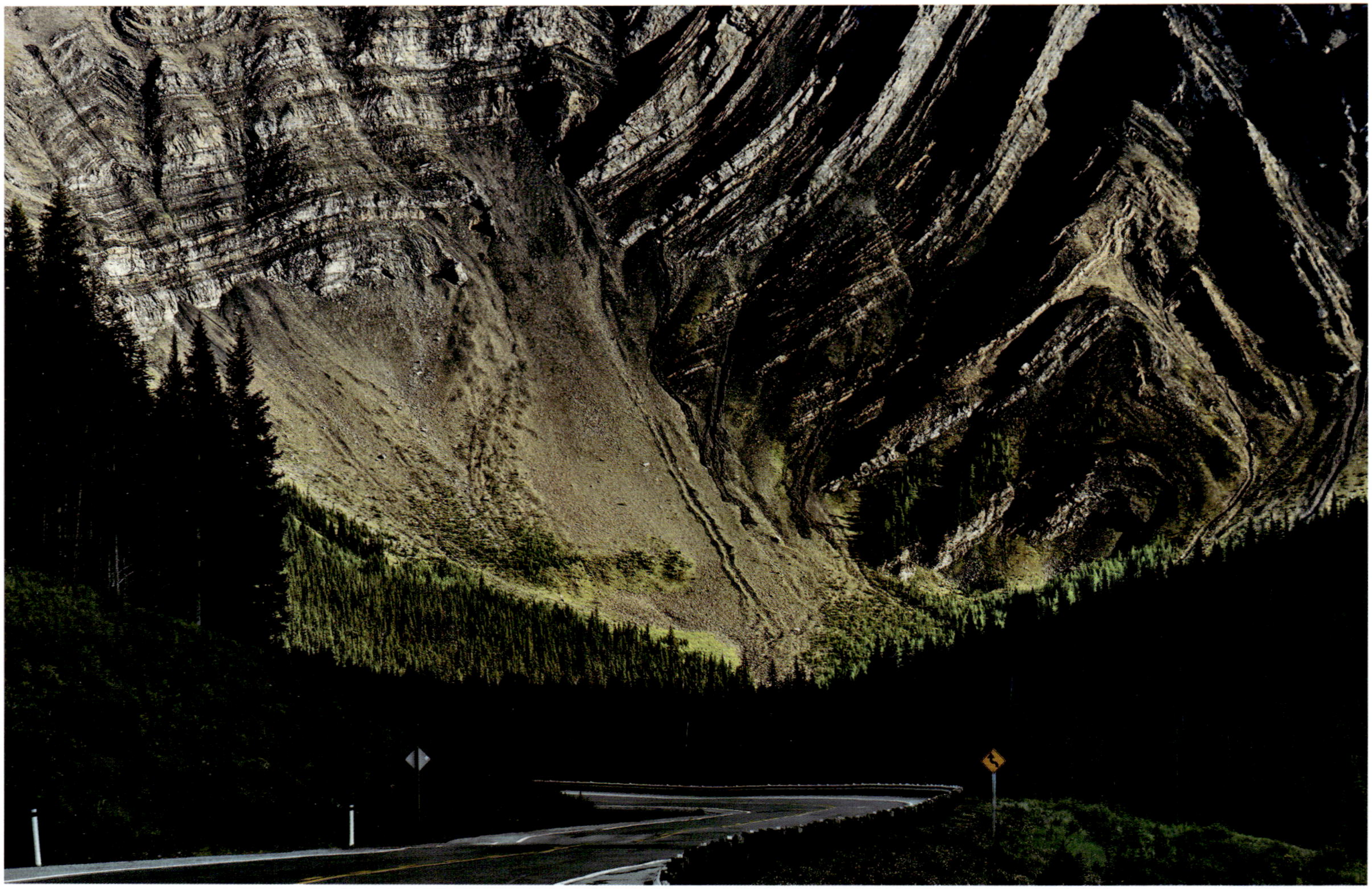

be followed by others who build something else. I use "landscape" loosely, to cover cities as well. I think this is fair in Canada where, in contrast to much of Europe and many parts of the United States, many cities are still ragged at the edges and never too far from what Canadians refer to as "the land." This was brought home to me last August while travelling through southern Manitoba. The air, even in the heart of Winnipeg, was full of the smell of stubble burning in the wheat fields to the south around Winkler and Morris (*Fields on Fire*, page 57).

Perhaps it's because Canada is still so young and raw as a settled, urban place that the tension created by the encroachment of the landscape on our living spaces could possibly make sense as a defining characteristic of the Canadian "sense of place." Somewhere in the shared Canadian memory there must be a recollection of efforts to hold onto a place to live in an unforgiving environment. I suspect that even Canadians born and raised in large cities may feel this. The painter and sculptor Harold Town in his introduction to *Canada with Love,* a book of Canadian photographs, wrote that this country has "no superb urban centres to soundproof us from the call of the wild." No matter how settled we may feel there is a part of us that understands the fragility, the tenuousness, of our position, and the magnitude of what we are up against.

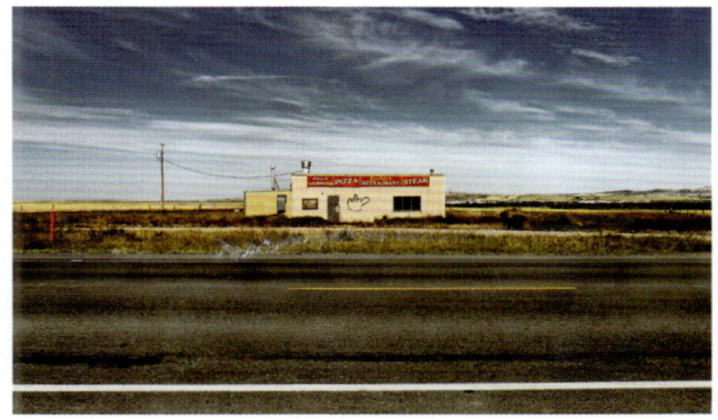

This idea accompanied me as I travelled around Canada. The connecting thread was the road, the ubiquitous sign of human presence in and movement across the landscape. Everywhere, in every image, the road is a reminder of the human urge—benevolent or malignant, as the case may be—to conquer, to overcome, to withstand, to appropriate, to build, to make a mark, to communicate, to carve out territory, and, above all, to get somewhere. Every road carries its own stories of

human ambition and frustration and longing and striving—stories that echo off the unhearing landscape, and that, like tire tracks in dust, fade and are overwritten.

In all of the travel required to take these photographs, I had the opportunity to see more of Canada than most people ever will. It was the experience of a lifetime. Like anyone embarking on such a journey, I carried with me a particular way of seeing the country. I don't know for certain whether this had anything to do with being Canadian, but I would like to think that it did. I keep coming back to the idea that what each one of us sees is filtered through our "essence." As I stood on, beside, and above roads all across Canada, I saw visions of a fragile and transient human presence in a landscape that was variously beautiful, harsh, magnificent, and desolate. Did I see things this way because I was born and raised in Canada, and, particularly, because I spent my childhood in the hard and unforgiving landscape of northwestern Ontario? I'm sure this is not the complete explanation; the vagaries of heredity and personality certainly had something to do with what I have produced here. On the other hand, I can't help but believe that the way of seeing captured in these photographs is somehow the product of a Canadian essence that opens our eyes to a shared sense of place.

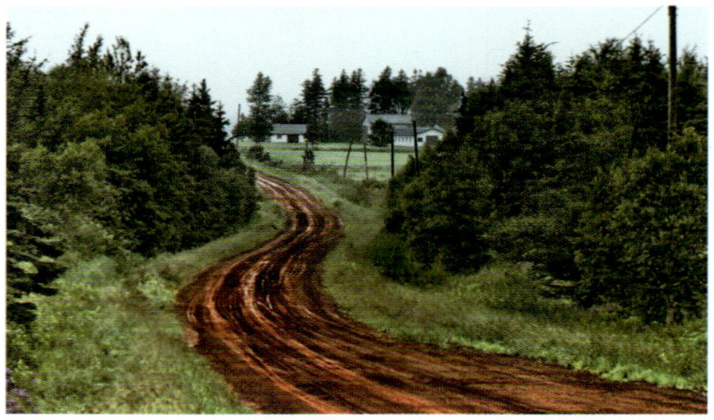

3

As I travelled around Canada in the summer of 2009, taking photographs for this book, I re-read Jack Kerouac's masterpiece of a novel, *On the Road.* In almost every way, the fictionalized story of his alcohol-soaked trips by car, bus, and boxcar, back and forth across the United States in the company of dissolute friends and strangers, had nothing in common with my sober, solitary travels in the relative

comfort of airplanes and rental cars. But we do share one thing—we were both searching for a way to feel and describe a connection with a land that is big and full of empty spaces, its soul stitched together by roads. Kerouac describes a journey down roads that by their nature never end because they carry people driven by an irresolvable tension. Kerouac himself, and the people he wrote about, had an insatiable hunger for movement and change; on the other hand, they were also seeking rootedness, connection, and identity. It's the human condition, plain and simple. We have no choice but to accept it and to "lean forward to the next crazy venture beneath the skies."

The notion of restlessness in competition with a yearning for rootedness impressed itself upon me with increasing force as I completed the work for *Roads*. You don't travel far through Canada without being reminded that from the sixteenth century onward this country's story is a tale written mostly by immigrants. The early settlers who came here were restless people, driven by some combination of ambition, desperation, greed, a thirst for adventure, and a simple desire for a "better life," to seek fortune and opportunity far from home in an unknown place. Today we are still—by necessity if nothing else—a highly mobile people. And roads are the surest sign not only of our restlessness but also of our expectation that we should and will always have the freedom to keep moving, to go as far as the roads will take us.

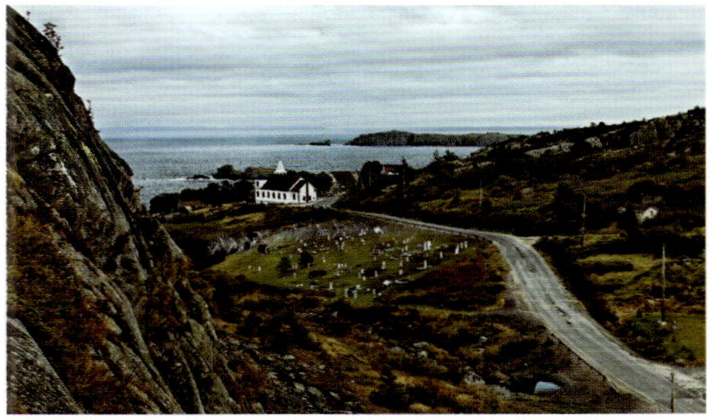

At the same time, strung out along roads across prairie, urban centres, and forgotten hamlets you see the signs—churches, community centres, and fairgrounds; town halls and bingo halls and movie theatres and bars—of our no-less-powerful desire for comforts to be found (so we hope) in community with others. You will see these tokens of our collective split personality in many of the photographs.

I couldn't help but be haunted by Kerouac's vision of the road as an emblem of human frailty. As he passes through North Platte, Nebraska, Kerouac's autobiographical narrator, Sal Paradise, observes the "false-front buildings. There were immense vistas of the plains beyond every sad street." The same description might apply to any number of prairie towns I passed through in Manitoba, Saskatchewan, and Alberta. If I have been able to portray nearly as vividly in photographs the sense of place that Kerouac created with those few words, then my effort will have been worth something.

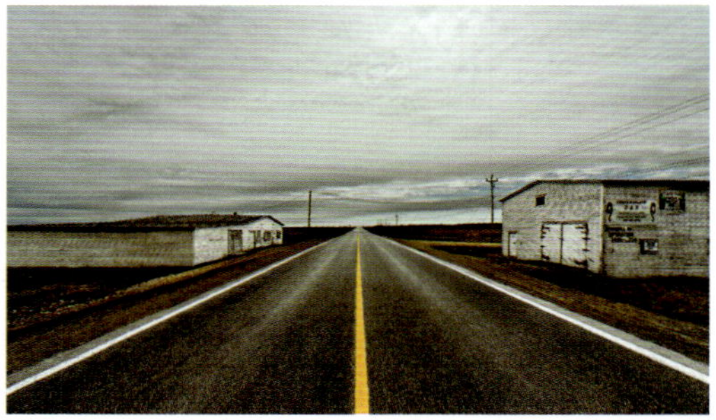

4

I want to offer a word of thanks. Undertaking a project of this nature—one that involves thousands of kilometres of travel by car and airplane—consumes money as well as time. I am grateful for the support I received to defray travel costs from the official tourism agencies of (in alphabetical order) Alberta, New Brunswick, Newfoundland, Nova Scotia, Ontario, Prince Edward Island, Quebec, Saskatchewan, and Yukon, as well as support received from the Canadian Tourism Commission. Their generous assistance lightened the burden considerably. I thank Fifth House for its decision to publish this book. And above all I thank my wife and daughters for their love and encouragement throughout.

Finally, a note about equipment. I shot most of the photographs in *Roads* with a Canon 1Ds Mark III digital camera. I used a variety of lenses that ranged in focal length from 12mm to 400mm.

—Mark Schacter

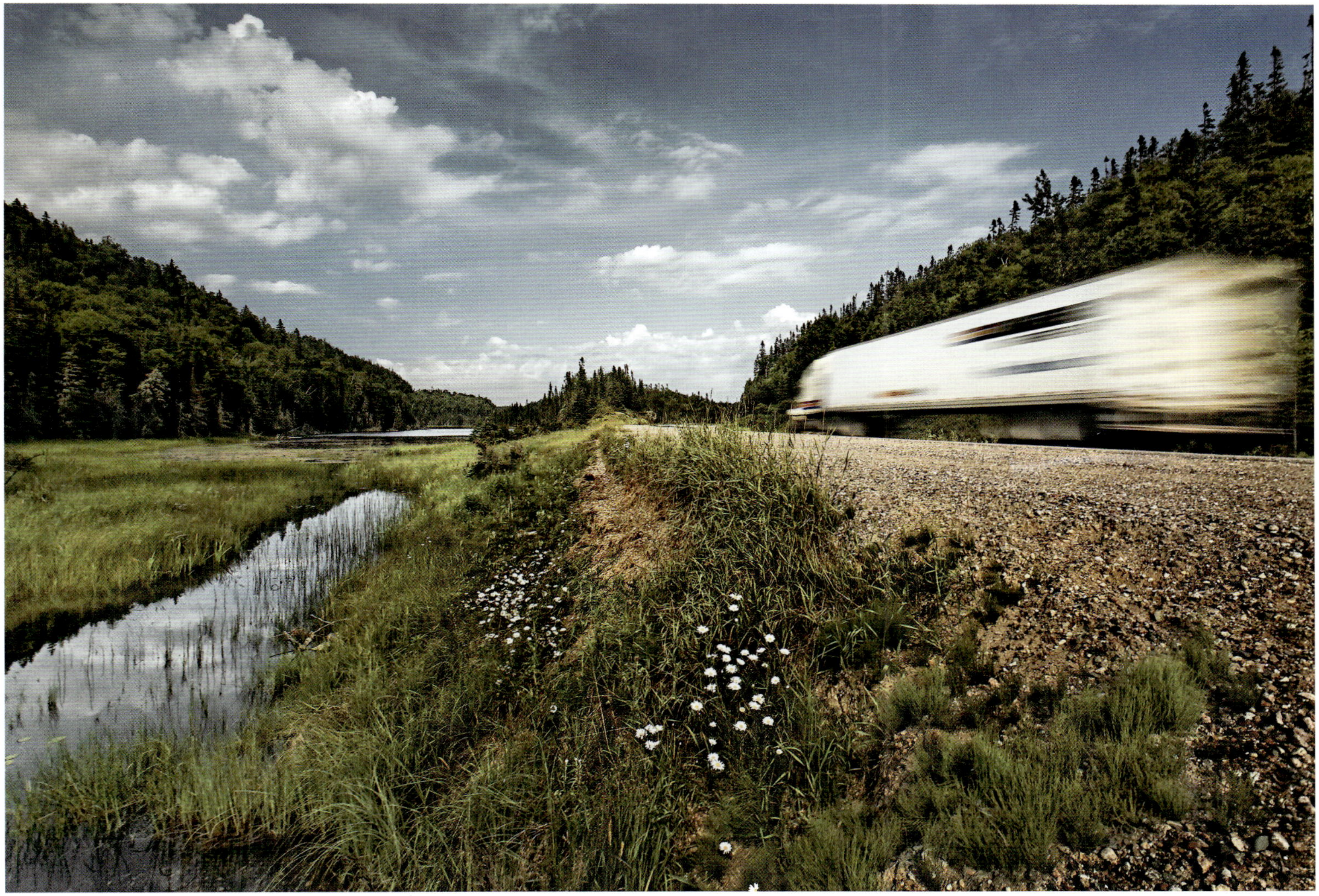

TransCanada Highway, near Terrace Bay ON

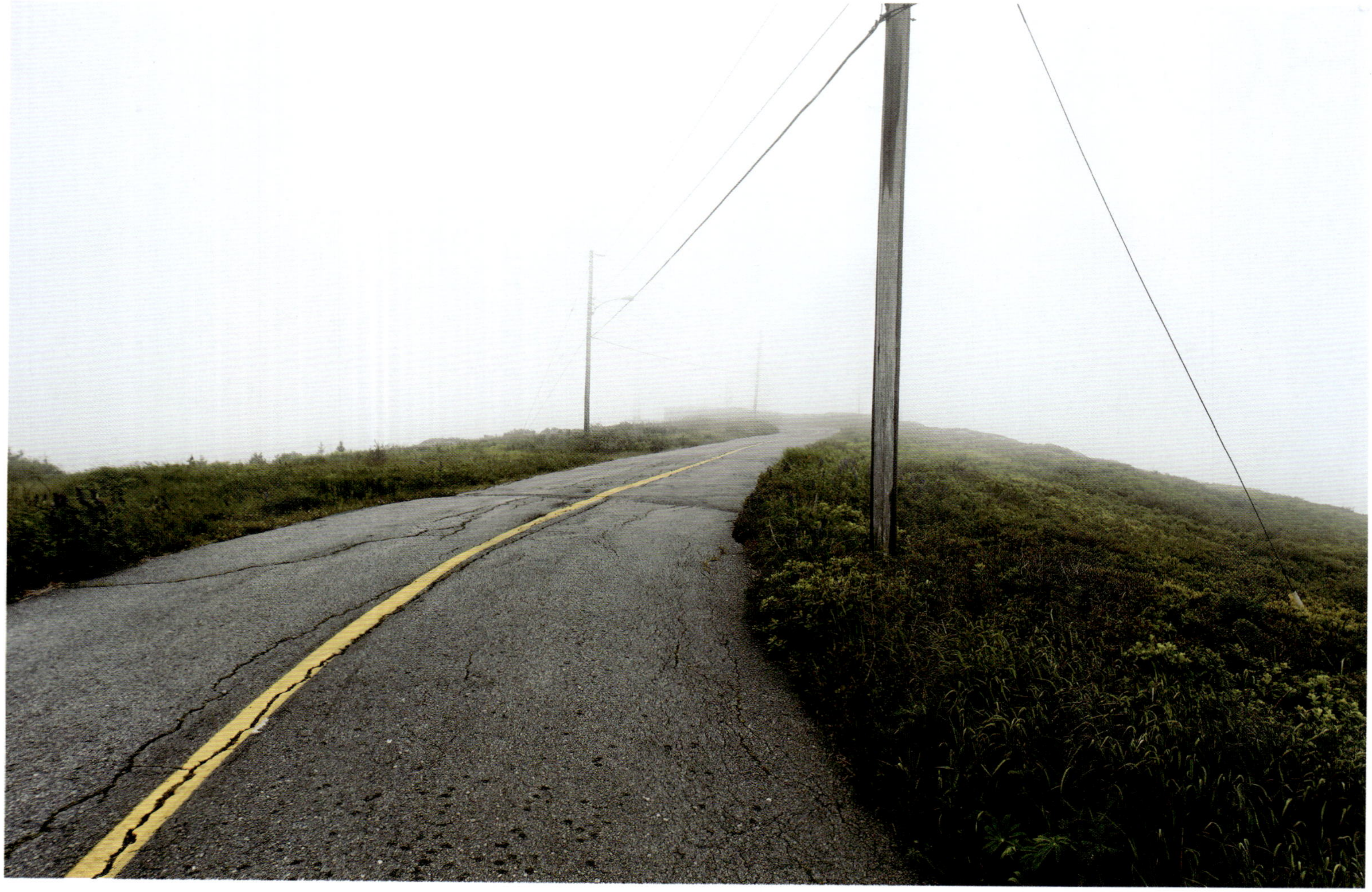

ROADS

ROADS

Mark Schacter

FIFTH HOUSE

PART 1

What You See = Where You're From

DURING THE COURSE OF GATHERING the material for *Roads*, I took about 15,000 photographs. That implies a lot of choices—choices to photograph certain things and not to photograph others. And among the 15,000 there were many—indeed the vast majority—that I decided not to use.

My choices were rooted in a particular way of seeing things. Gotthard Schuh, a Swiss photographer, said that "Each person photographs only what he sees, and each person sees only what accords with his own essence." That "essence," in my case, is a product of where I have come from: a city in northwestern Ontario, called Thunder Bay. I was born there and left in 1973 when I was sixteen. I don't miss Thunder Bay and have never had any desire to return there to live. But the fact that I was born and raised in Thunder Bay is the first thing that comes to mind when I search for an explanation of what interests me as a photographer.

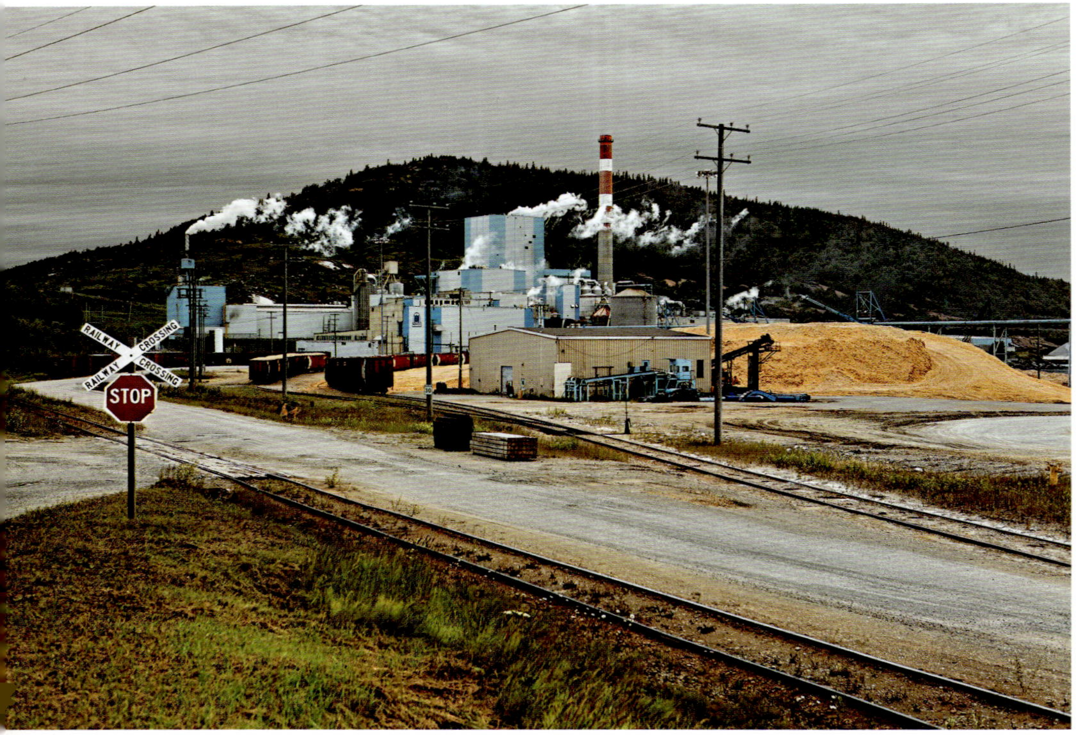

Paper Mill, Marathon ON

In *On the Road* Jack Kerouac wrote about "crackerbox towns with a sea for the end of every street," flimsy looking places that appear to be hanging on by their fingernails to the edge of nothing. Thunder Bay fits that description well. It sits on the western edge of Lake Superior, an inland, freshwater sea—the largest freshwater lake on Earth. In every other direction there is an undulating carpet of forests, rivers, lakes, and hills carved by time and glaciers. My habitual response to people who ask me what Thunder Bay "is like" is to say that it is an unattractive city in a magnificent landscape.

Though Thunder Bay is not tiny—the population today is about 120,000—it looks raw and unfinished in the way so many Canadian towns and cities do. And it is remote. The nearest place of any statistical consequence is three hundred kilometres away, across the US border in Minnesota. If you want to stay in Canada to find a city of significant size, then you have to go even farther—about seven hundred kilometres east to Sault Ste. Marie or the same distance, west, to Winnipeg.

Thunder Bay began with the fur trade and later became a hub for logging, paper mills, mining, and the trans-shipment of grain from western Canada to the world. Now these sources of industry and wealth are on the wane. Europeans first settled the area in the seventeenth century, but today Thunder Bay still looks as if it were only recently hacked out of the wilderness. It has the air of a place cobbled together without an obvious plan—like a house with additions and extensions bolted on as the need arose. You get the feeling that if the people whose job it is to watch over Thunder Bay were to drop their guard for a moment too long, the wilderness would quickly conquer it, leaving few traces of human presence.

My father was born in Thunder Bay in 1918. I remember being about twelve or thirteen when he gave me an old Kodak twin-lens reflex

Ontario Highway 26, Looking West, Craigleith ON

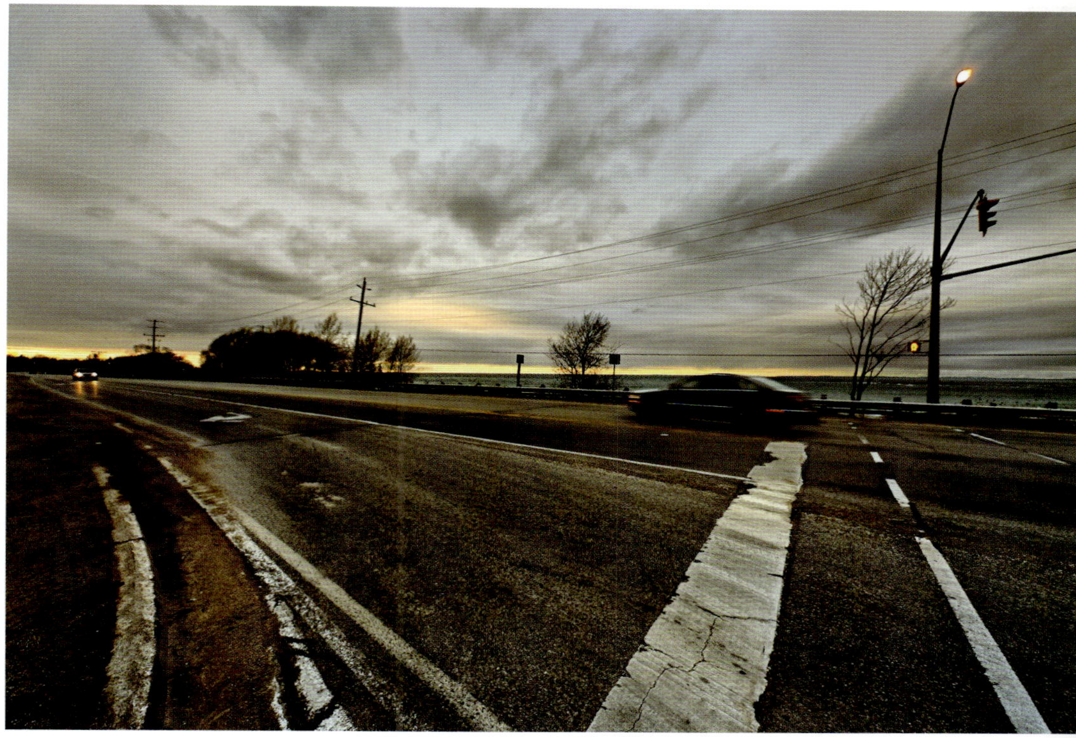

camera that his brother (who drowned in a boating accident before I was born) had bought. It came with a stout leather case and took 620-size medium format film. I remember one of the first photographs I shot with it: the subject was an old chair that had been tossed into our backyard. It was winter and the chair was stuck, tilted at a crazy angle, in the top of a snowdrift.

My father's father came to North America from Eastern Europe in the early 1900s. He lived in New York City before eventually making his way to Thunder Bay. He was a religious Jew who earned a living as, among other things, a junk peddler. On that shaky foundation, and out of the poverty of the Great Depression, my father founded and operated a successful scrap iron and metal business in Thunder Bay. He was often on the road, visiting the mines and mills and hydro-electric stations and other places from which he bought scrap steel and non-ferrous metal, which he then sold to steel mills and foundries that melted it down and made it new.

At some point my father began taking me with him on the road. I have memories from the sixties and early seventies of my father at the wheel, cigar in mouth, as we rolled along the single lane of the TransCanada Highway through the thinly populated reaches of northwestern Ontario. The place names still have a kind of magical resonance for me: Red Rock, Nipigon (*Front Street,* p. 18; *Road and Rail Bridges over Nipigon River,* p. 91), Schreiber, Rossport, Terrace Bay, Marathon (*Paper Mill,* p. 16), Beardmore, Jellicoe, Geraldton, Longlac.

The sensations of those road trips were power and speed—my dad's humming two-seater sports car whizzing through deep rock cuts blasted out of the Precambrian Shield. At the same time there was a feeling of insignificance—two people alone in a hard, empty, and indifferent landscape. The little towns looked makeshift and vulnerable: clearings scraped from the

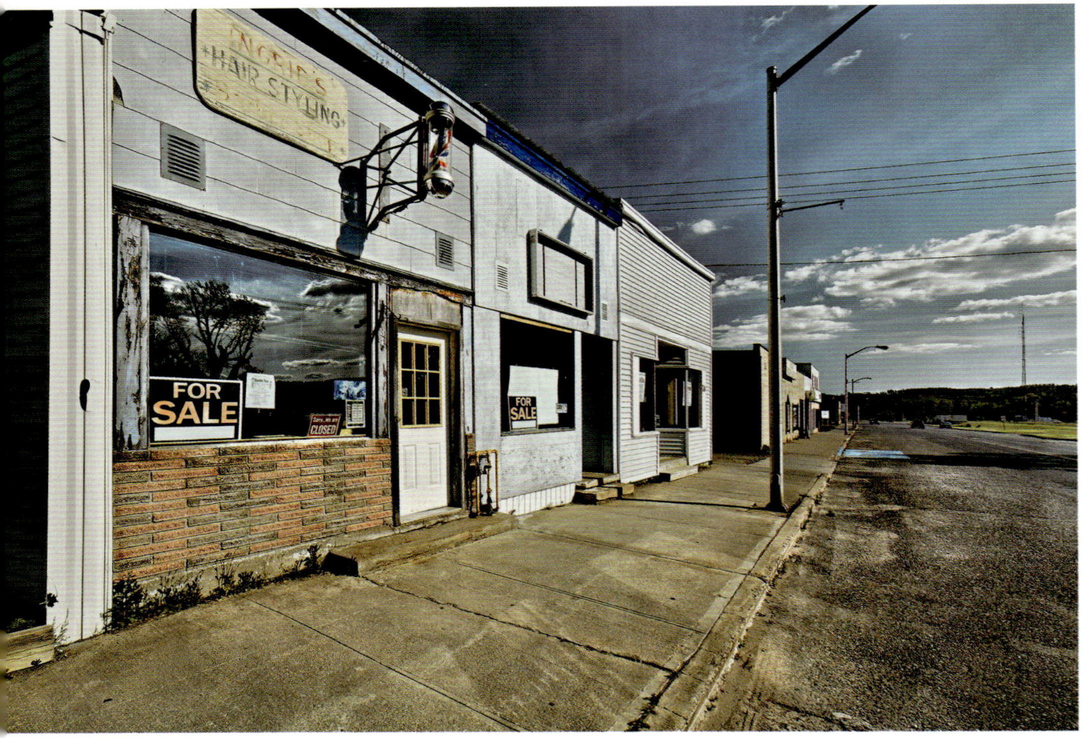

Front Street, Nipigon ON

rock and forest only (so it seemed) to be swallowed up again when, eventually, the gold and copper were mined out, the forests depleted, or the commodity markets turned sour.

I loved the emptiness of the highway, the centre line fluctuating from solid to broken according to the lay of the land. I loved the highway signs, white letters and numbers on a green background, announcing places and distances. I loved the mystery of lakes and rivers glimpsed through the darkness of the boreal forest. I liked the look and feeling of the sad and solitary little towns, and the acrid sulphur smell of the pulp and paper mills where many of their residents worked.

In the summer of 2007, I went back there to spend a week taking photographs along the north shore of Lake Superior, travelling from Thunder Bay in the east to Marathon in the west, and back again. I pulled up to the gate of the paper mill in Red Rock—a regular stop on the outings with my father—only to find it had shut down a few months earlier (*Paper Mill—Shut Down*, p 115). Nearby was Nipigon, another place dependent on the forest, and it looked like a ghost town. In Thunder Bay itself, two of the city's paper mills were closed; its port, once the largest grain handling facility in the world, was relatively quiet, and a swath of land once used for grain handling had been converted into a waterfront park. The massive iron-ore dock down at the lake shore, built to load Great Lakes freighters with hematite carried by rail from a mine two hundred kilometres to the west in Atikokan, was derelict, long since stripped of its spidery ore chutes and storage bins.

Canada is filled with such places—towns and cities that exist because there once was, or still is, ore to be mined, trees to be cut, oil to be pumped, grain to be harvested, cattle to be slaughtered, and fish to be caught. For all the apparent sophistication of places like Vancouver, Toronto, Montreal, or Quebec City, we remain, more or less, a nation of mines and smelters; oil rigs and refineries; logging camps, sawmills, and paper mills; grain elevators, stock pens, and fishing boats. Neither the leather-lined luxury cars plying the streets of urban downtowns, nor the sleek, oversized pickup trucks creating miniature dust storms as they speed along arrow-straight prairie roads, can outrun the old Canadian refrain of extraction and exhaustion. That's how it is where I come from, and roads never fail to remind me of it.

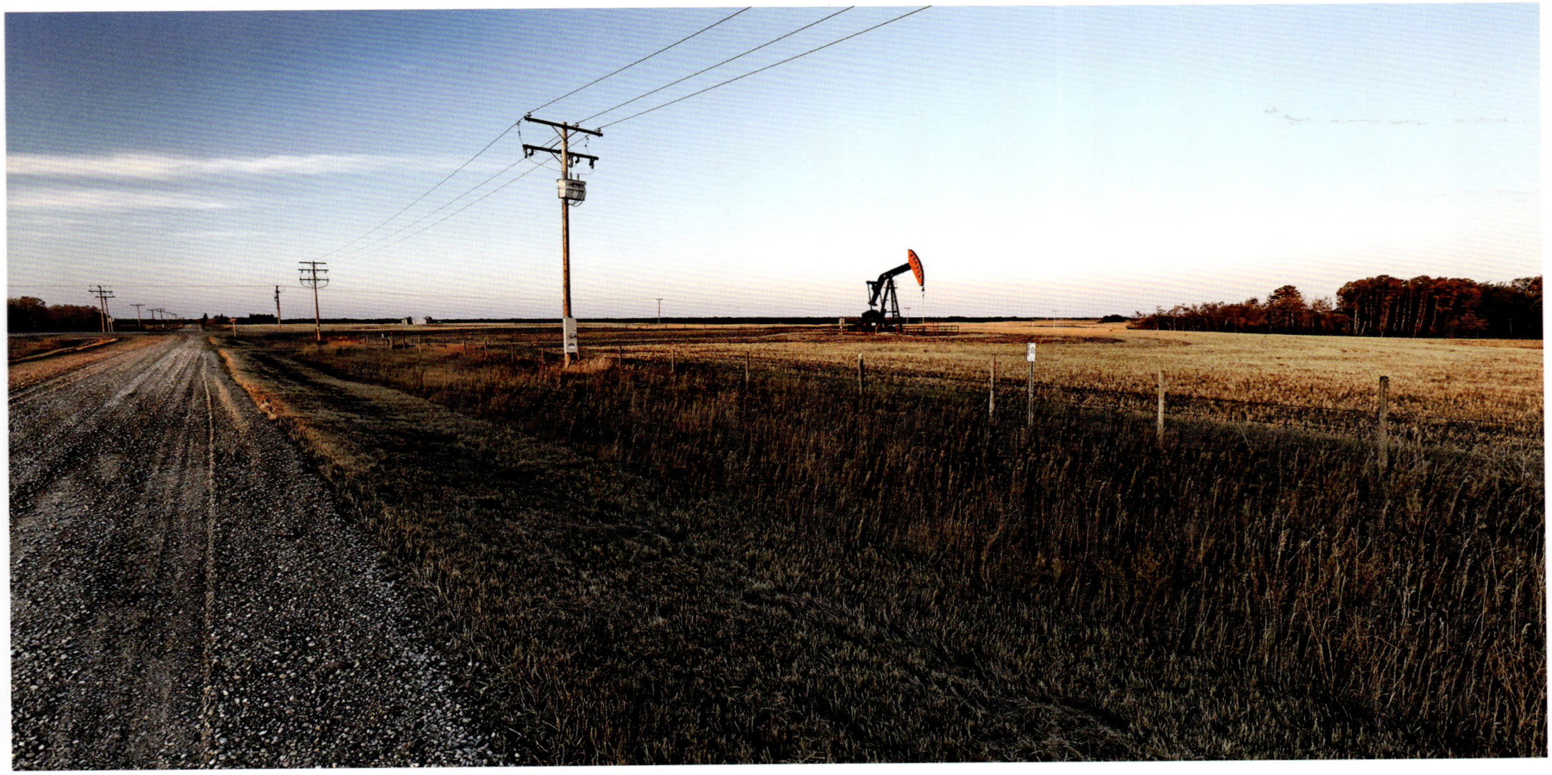

Oil Well, near Manor SK

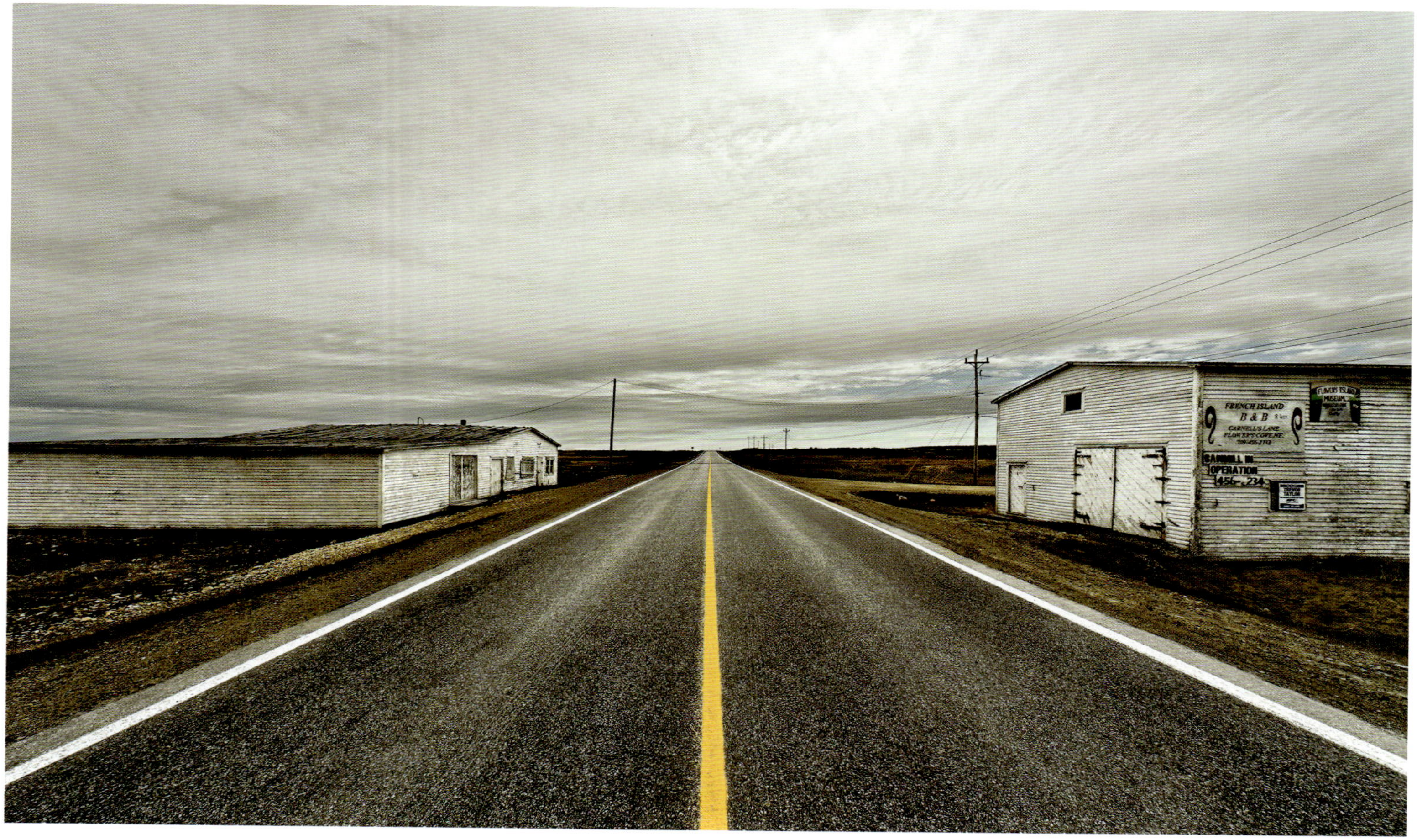

Road out of Anchor Point, Anchor Point NL

PART 2

On Being Lucky

> Somewhere along the line I knew there'd be girls, visions, everything; somewhere along the line the pearl would be handed to me. —Jack Kerouac, *On the Road*

THERE ARE ABOUT 1.4 MILLION KILOMETRES of roads in Canada. One million, four hundred thousand kilometres is almost four times the distance from Earth to the moon; more prosaically, it's about one hundred round trips by car and ferry between St. John's and Vancouver. This simple, brute fact of distance begs some obvious questions. Over which roads would I travel? How would I *decide* where to go? What would my plan be?

Many people asked me these questions, but I wasn't able to provide them with a satisfying answer. The sum total of my plan was that I

- would cover every province and territory (except Nunavut, for logistical reasons) as extensively as my time and resources would permit

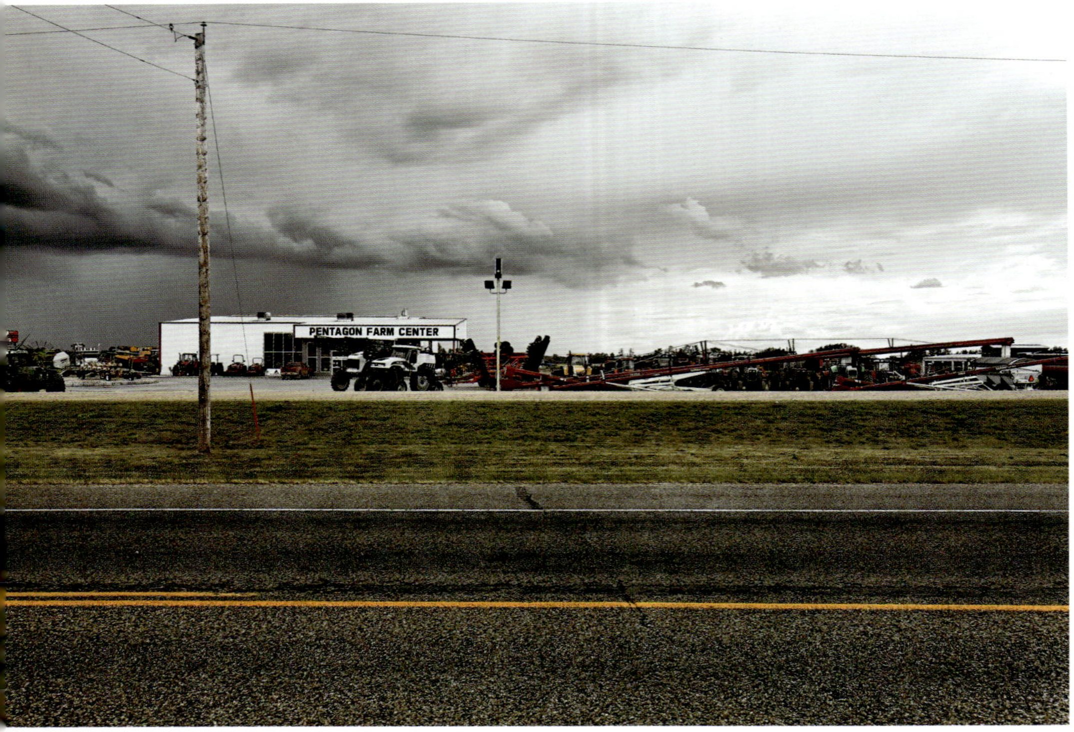

Prairie Storm, Westlock AB

- would include a road somewhere in every photograph (not necessarily in a prominent position, but somewhere)
- wanted a balance of landscape, rural/small town, and urban photographs
- would (for reasons related to ease of travel and a preference for keeping my hands warm) do a relatively small amount of the work in winter.

Put simply, my plan depended to a large extent on being lucky. And being lucky, I suggest, is something that can, at least in part, be planned for.

Faced with a virtually infinite supply of potential road scenes, I had little notion of where I would go once I had touched down at an airport and got behind the wheel of my rental car. The limiting factors were time—the number of days I allocated to each part of the country (a calculation based roughly on my travel budget) and hours of daylight (a function of latitude and time of year)—and weather. A constant driving rain would put me out of action, not because it would damage my equipment (my camera and lenses can withstand most rainy conditions) but because water spatters on your lens make for bad photographs.

Aside from those few fundamental constraints, everything depended on me keeping my eyes open: being alert to scenes with the potential to become interesting photographs; being intensely aware of my surroundings as I drove the car (while still paying attention to the task of driving!). And often it meant taking the time to get out of the car and walk around for no immediate reason other than to see what I might find.

Luck … circumstance … being in the right place at the right time … these were part of the story behind many of the photographs in this book. But luck—I'm convinced of this—happens in a context. The trick was to put myself in a position where it would be easy for luck to find me.

Where "lucky" photographs are concerned, there are at least three things to consider.

Corner of 4th and Main, Cardston AB

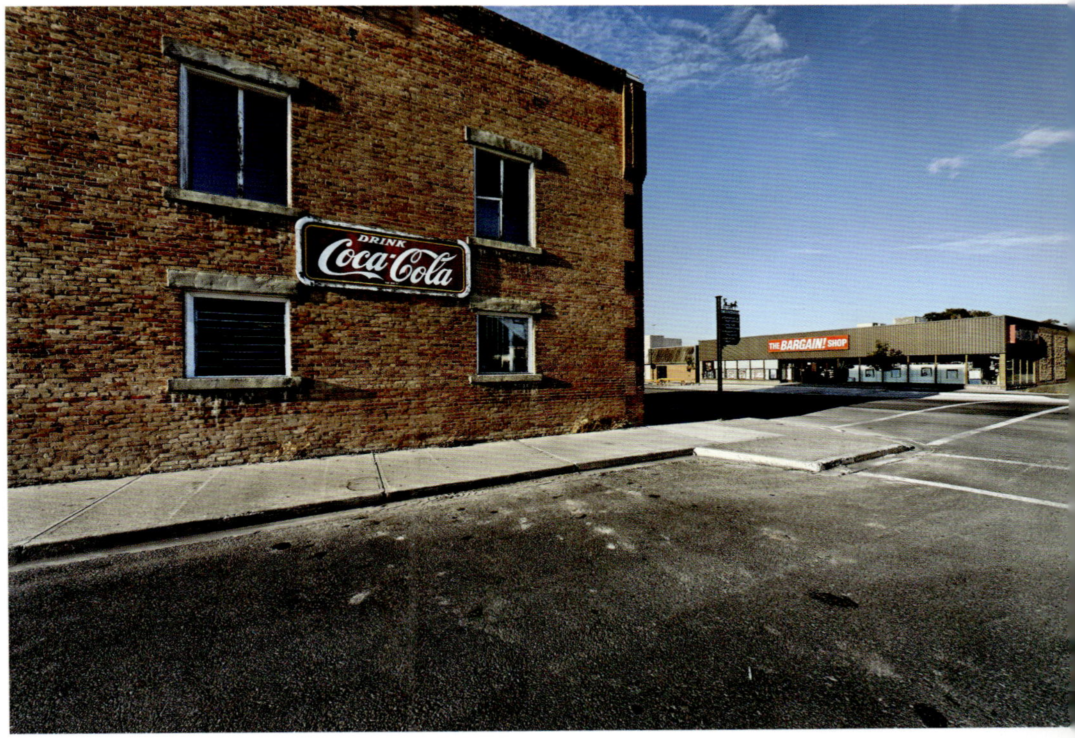

- First, there must be a photographic opportunity, a thing to be photographed.
- Second, I must somehow happen to be present when the photographic opportunity materializes in a particular place, at a particular time.
- Third, I have to *recognize* that an opportunity is staring me in the face, and be ready to capture it. (Henri Cartier-Bresson, the French photographer, spoke about his pursuit of the "decisive moment"—an instant in time and space when a composition presents itself. Cartier-Bresson once told an interviewer that there is "a creative fraction of a second when you are taking a picture. Your eye must see a composition or an expression that life itself offers you, and you must know with intuition when to click the camera. That is the moment the photographer is creative. Oop! The Moment! Once you miss it, it is gone forever.")

Of these three considerations, the first is out of my control. The same cannot be said for the other two. Though it sounds like happenstance, being in the right place at the right time to seize an opportunity was, in the case of this photographic journey, at least partly due to my own actions. And the third element was completely my doing—it was about my *eye*, my decisions. And so if I were to give each of these factors equal weight—the event; the being there; and the recognition of the opportunity—I would have to conclude that I was responsible for somewhere between one-third and two-thirds of the "luck" in a lucky photograph.

Events on the afternoon of August 18, 2009, provided one example. A storm was blowing across the prairie north of Edmonton, near Westlock, Alberta. Viewed from the east side of Alberta Highway 44, the clouds looked like they were about to swallow up the premises of a farm equipment dealership (*Prairie Storm,* p. 22). This was the first ingredient of the

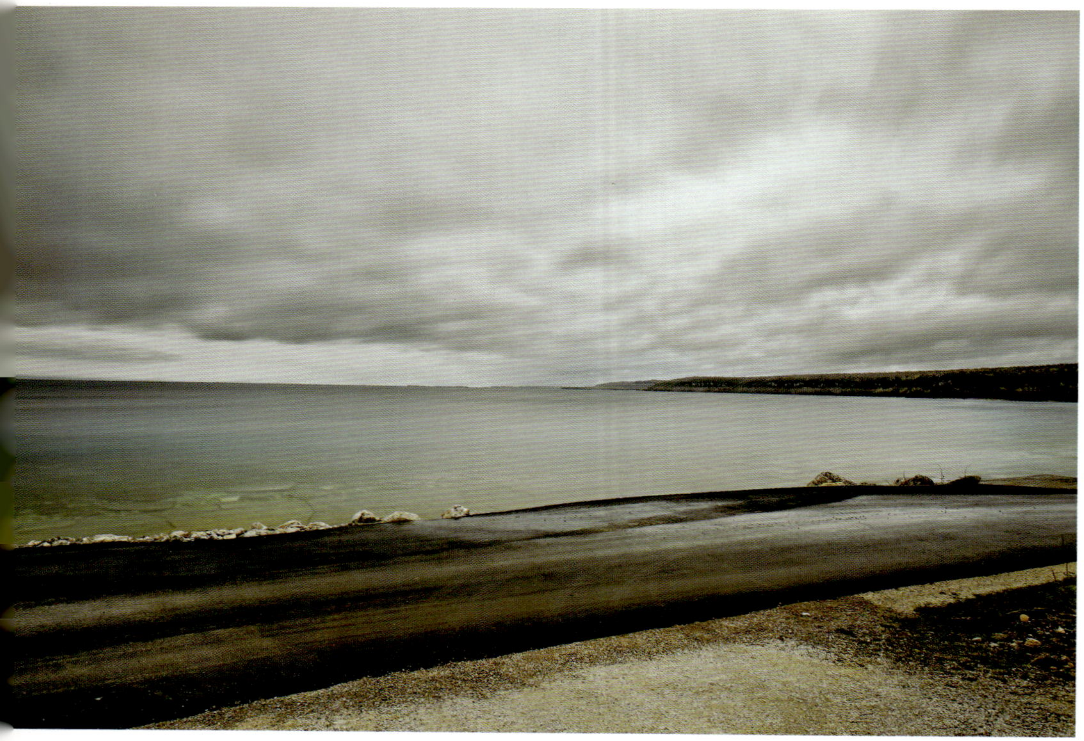

Barrow Bay, Barrow Bay ON

luck cocktail—an event that created a photographic opportunity. Several hours earlier, I had been belted into the aisle seat of a flight from Ottawa to Edmonton. The boy seated at the window told me he lived on a farm near Westlock. Later, about to pull out of the car rental parking lot at the Edmonton airport, I decided to drive north toward Westlock, for no reason other than that the boy had mentioned the place. There's the second ingredient—the thing that brought me to the place of the photographic opportunity was my decision (and it *was* a decision) to take a whimsical approach to planning my route. And then finally—the third ingredient—as I was driving up Highway 44, I recognized out of the corner of my eye that the juxtaposition of angry clouds, prairie, and the white building on the left-hand side of the car might make for an interesting photograph.

Ten days later, at 7:30 pm, a horse began to drink from a tiny pond (*Horse Drinking,* p. 148) at the junction of Alberta Highway 5 and Range Road 264 in the hamlet of Leavitt (ingredient no.1–the event). A short while earlier I was leaving Cardston, where I had stopped to take photographs (*Corner of 4th and Main,* p. 23). I was making my way back to Pincher Creek to the farmhouse bed and breakfast where I was staying. The most direct route (appealing, because it was getting late, and I was tired and hungry), was to go north out of Cardston and then west. But I detoured to the south because I was attracted by a name on the map—Mountain View (ingredient no. 2—the decision that brought me to the place of opportunity). On my way there, as I passed through Leavitt, something (I can't remember what) caught my eye on the left-hand side of the road, so I looked for the first possible place to turn the car around. This brought me to the little pond, toward which the horse was slowly moving. I quickly put aside whatever it was that had caused me to stop in the first place (ingredient no. 3—recognition of the opportunity), and took a photograph of the horse instead.

Bruno Hotel, Bruno SK

And I recall another photograph where the act of quenching a thirst was the thing that caught my eye. Shortly before 1:00 pm on October 17, 2009, a woman paused in the sunshine outside the Bruno Hotel (*Bruno Hotel,* right) at the corner of Main Street and Railway Ave. North in Bruno, Saskatchewan (ingredient no. 1). A few minutes before, I had decided to park my car on Railway Ave. North so I could wander around the village (ingredient no. 2). It didn't take long for my eye to be drawn to the hard shadows and simple lines of the Bruno Hotel. I positioned myself on the other side of Main Street and began to take photographs. And then the woman walked into the photograph, paused in the nook at the corner of the building, and raised the drink to her lips—and I knew that *this* was the image I wanted (ingredient no. 3).

In short, I had some luck, and I was ready for it.

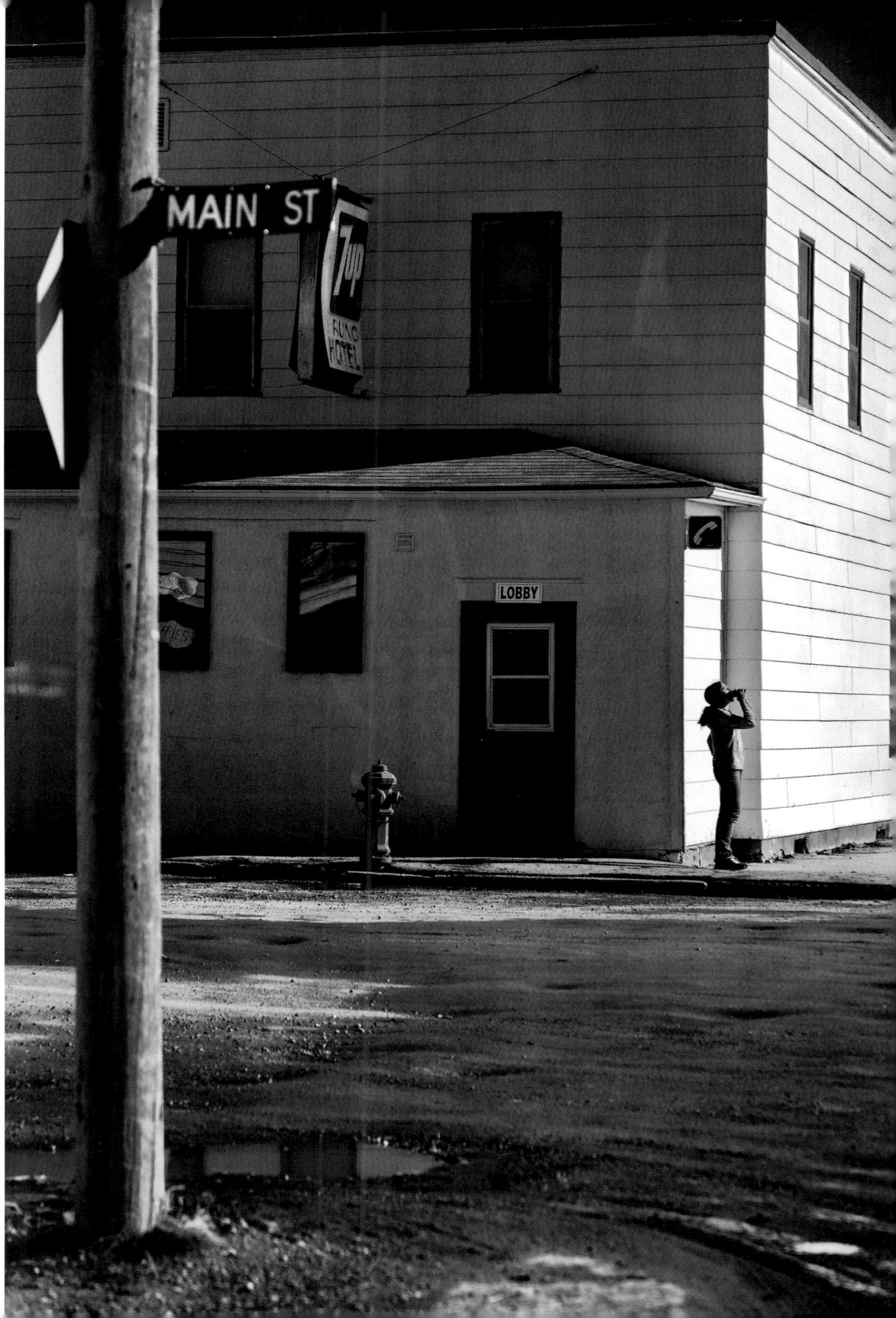

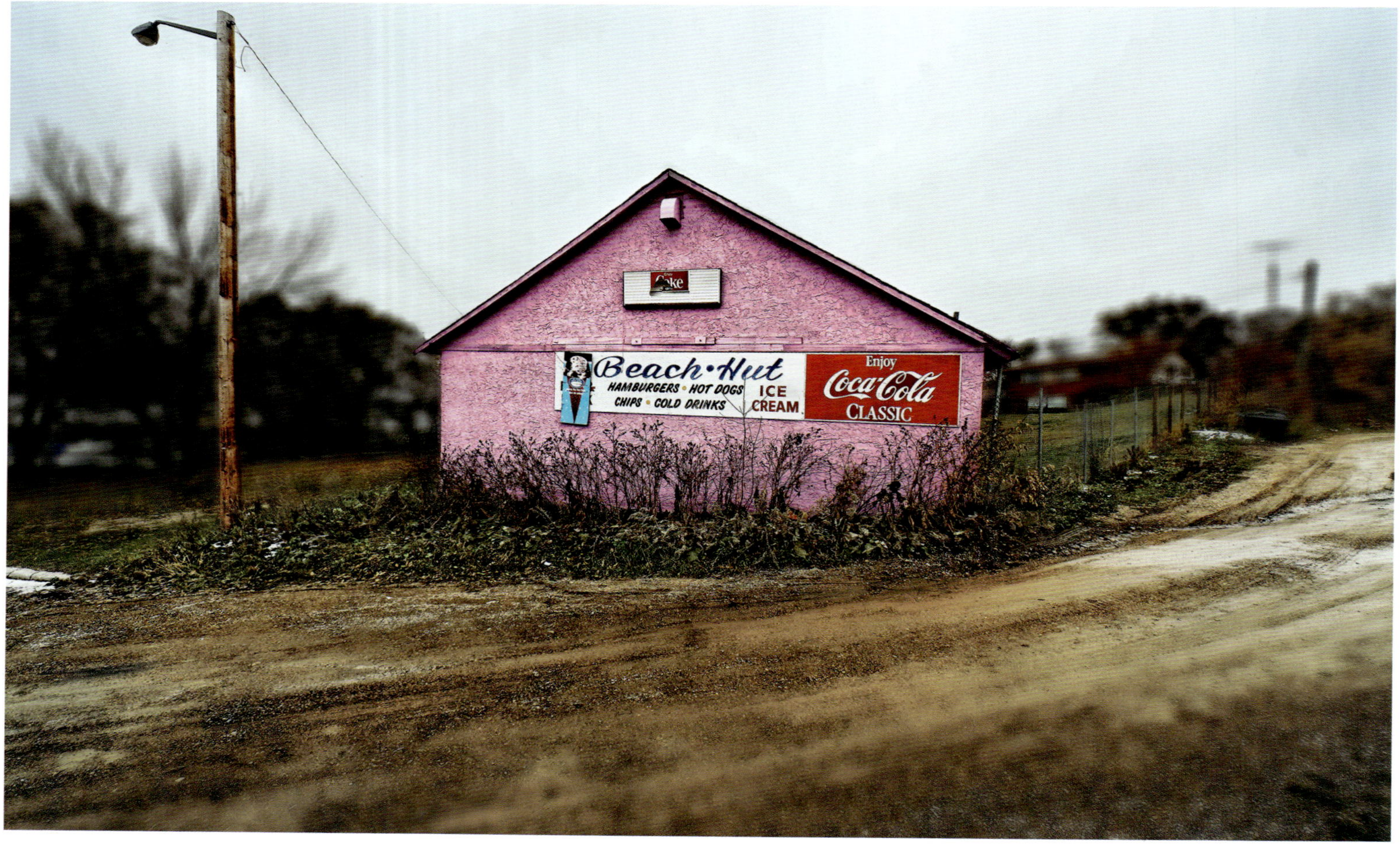

Beach Hut, Regina Beach SK

PART 3

But is it "Real"?

HAVING BEEN BORN AND RAISED in northwestern Ontario has left me especially sensitive to the idea of people isolated and small in a big and harsh landscape; of carving out a place to live and prosper in that landscape; and of the impermanence and flimsiness of human endeavor relative to the landscape. It has also given me a fondness for industrial scenes—mills, mines, ports—which some regard as a blight on the landscape but which I find powerful and beautiful. The photographs in this book attempt to express these ideas and the emotions that accompany them. I want to convey a sense of *place*; a sense of what it feels like, to me, to be somewhere.

The work is obviously subjective because it's an expression of my particular way of seeing. I emphasize this point because the idea of subjectivity and the practice of photography have not always coexisted comfortably. There is a special kind of discussion about the proper relationship of photography to the "real world"—a discussion that doesn't happen in relation to the other visual arts. We have no expectation, for example, that a "good" painter or sculptor will necessarily produce work that is an objectively accurate representation of reality. This is not always so with respect to photography.

Expectations of "good" photography (when judged as "art" rather than a form of documentation, such as photojournalism) have long been, and continue to be, debated. Unlike the hand of the painter or sculptor, a camera is a recording machine capable of making precise copies of whatever stands in front of it. In the minds of many this leads to the view that producing accurate representations is the proper purpose of photography. (The extent to which this is taken for granted is seen in common expressions such as "photographic memory" and "snapshot of the current situation.") According to this understanding, photographs that distort reality are misrepresentations; they amount to a form of cheating.

The debate about the photographer's role is as old as photography itself, but it has taken on renewed significance in recent years because of the omnipresence of digital photography. Even people who have only a casual understanding of photography know that photographers who use digital equipment can alter their images with software. Anyone with a camera, a computer, and a minimum amount of technical savvy now has the capacity (so the argument goes) to turn a sow's ear into a silk purse.

This point of view has given rise to doubts about the veracity of digital images. I am often asked the following two questions: first, "do you shoot with film or a digital camera?"; when I answer "digital" (most of the photographs in this book were shot with a Canon 1Ds Mark III) the inevitable follow-up question is, "have you manipulated your photographs?"

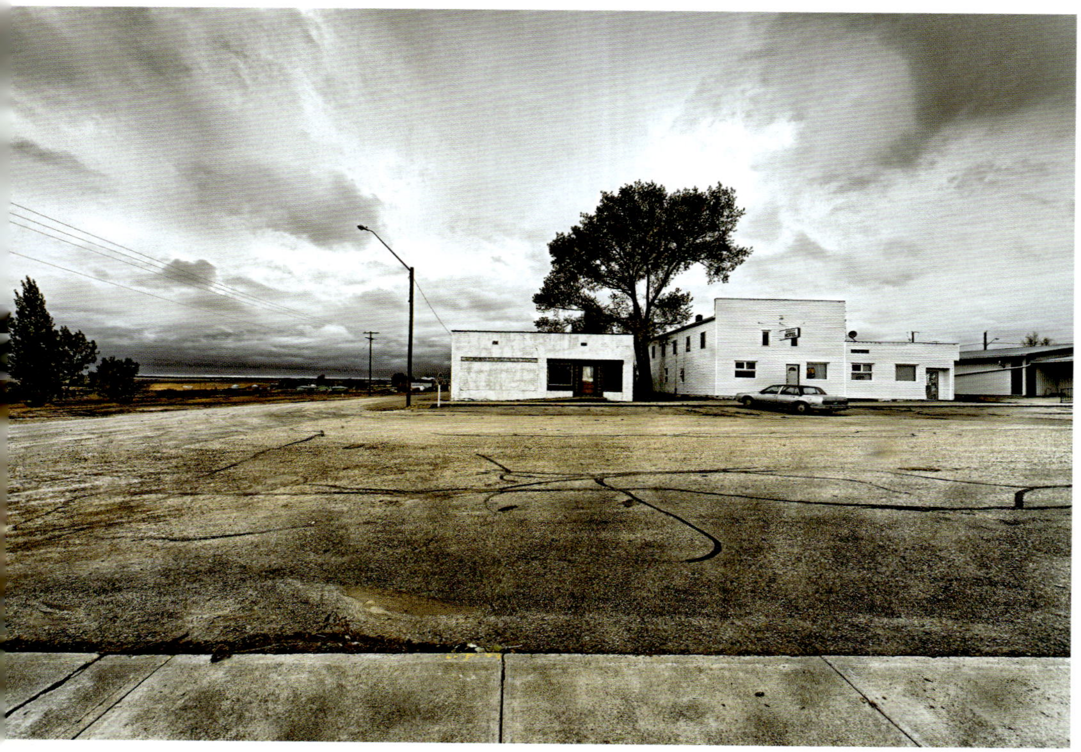

Beechy Hotel, Beechy SK

The answer to the second question is *yes*. All of the photographs in this book have been processed with digital editing software. The adjustments have affected, among other things, color tone, contrast, highlight detail, and shadow detail. My objective was to go for maximum impact; that is, to do the best possible job of expressing the ideas and emotions I wanted the images to convey. This editing work is an essential part of the creative process—at least as important as anything I do before shooting the photograph.

As I interpret it, the assumptions behind the two questions are that (a) photographers who use film (as opposed to digital) equipment do not manipulate their images or, at most, are much less manipulative than digital photographers; and that (b) a photograph should not be manipulated because it should capture the world "as it really is." In my view the first assumption is factually incorrect, and both assumptions are beside the point.

Long before the age of digital photography, Ansel Adams, perhaps the best-known of all landscape photographers, said, "You don't take a photograph. You make it." Photographers, whether they use film or digital equipment, have always made deliberate choices that affect how the world in front of the lens will finally appear as a finished photograph. The decision-making begins even before the photograph is taken.

On the most basic level, a camera records the world within the boundaries of a rigid and artificial frame—a 2 x 3 frame in the case of the most widely used hand-held cameras. This bears no resemblance to how we see the world with the naked eye. Every photographer is therefore required, right from the beginning of the image-making process, to *manipulate*: to make a decision about what to include in the frame and what to leave out.

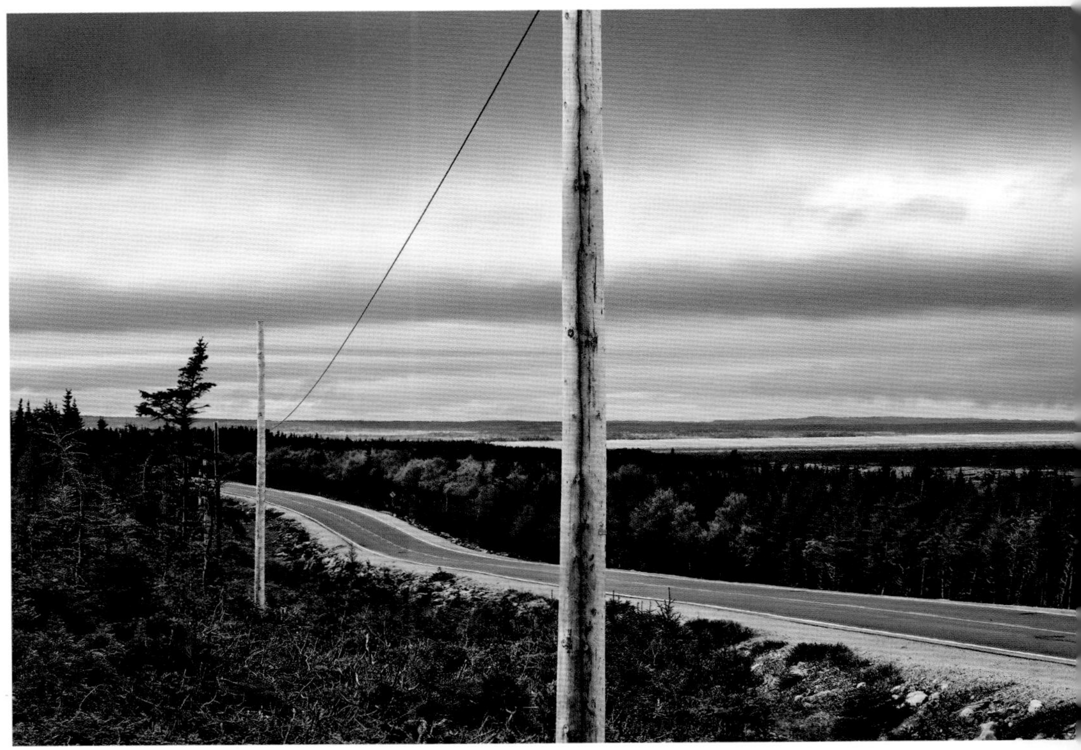

Highway 430 near Barr'd Harbour, Barr'd Harbour NL

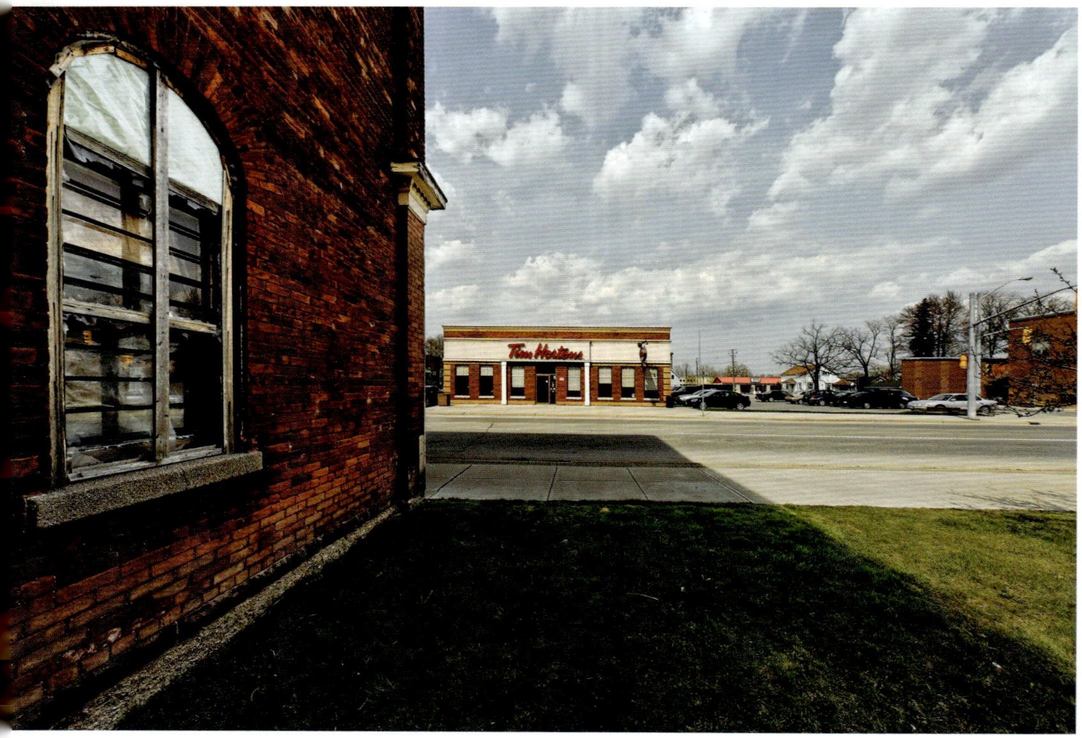

Tim Hortons, Petrolia ON

Other manipulative decisions abound. The film photographer decides on the type of film to be used. A fundamental choice is colour versus black and white. There is a popular perception that black-and-white photographs are somehow more authentically "artistic" than colour, but this perception seems to contradict the popular view that excessive manipulation is to be avoided. The real world, after all, is not black and white.

Further choices about film types are also critical to the appearance of the final image. Some films are grainier than others, giving the printed photograph a "rougher" or "smoother" appearance. Slide film produces a different look than print film. Some film types are more contrasty than others, yielding different ranges of grey values between the whitest white and the blackest black; they also respond differently to given wavelengths of light, so that, for example, the red of a stop sign might be rendered as dark grey by one black-and-white film and light grey by another. Similarly, colour films are distinguished by their unique colour temperatures—some will yield a warm (reddish or yellowish) colour cast, while others are colder (biased toward bluish tones), and others are more neutral.

Lighting also plays a role. A photographer's studio may have artificial lighting equipment designed to create a particular look for the finished image. Ironically, the objective of the artificial lighting may be to make things "look natural." Photographers working outdoors will also sometimes use artificial lighting (in addition to natural light) to enhance a scene.

And then there are the decisions the photographer makes at the moment the photo is taken, each of which affects how the camera records reality—shutter speed, aperture of the lens, the lens to be used (macro, wide angle, standard, zoom), whether to put a filter on the lens, whether to hand-hold the camera or put it on a tripod, etc.

Roadside Memorial, near St. Sixte QC

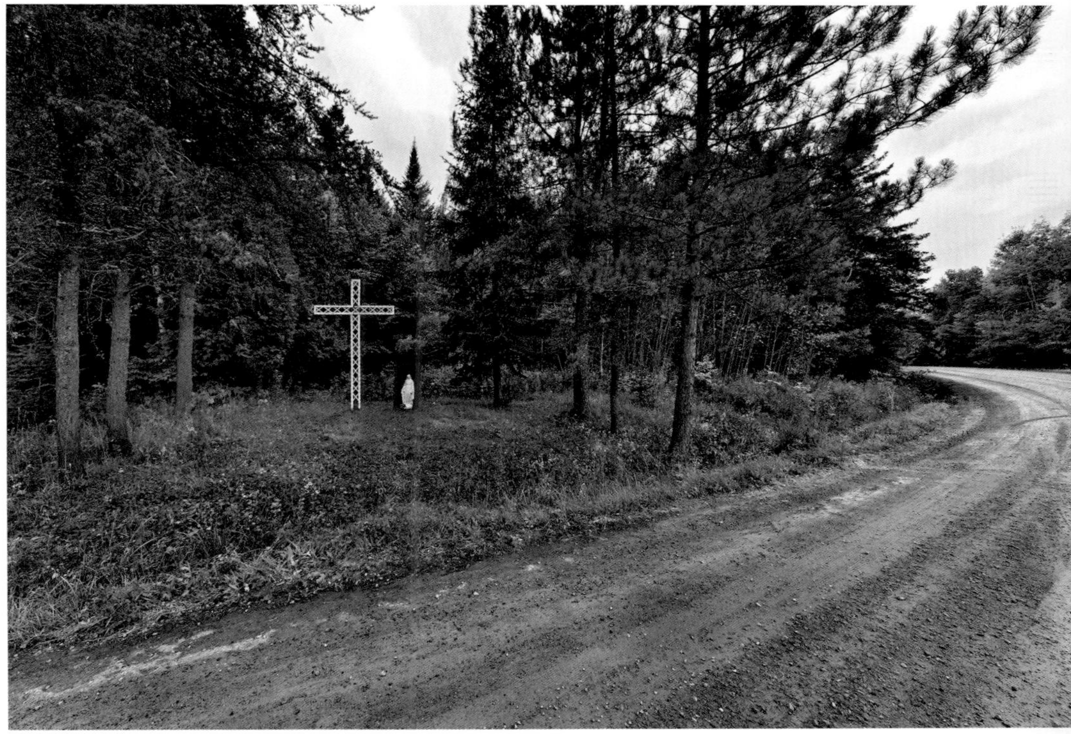

What I've just described is only the *first part* of the manipulative process: the part *before* the photograph is taken. I think the popular view is that the role of the film photographer ends at this point, while the digital photographer can go on to play with the image on the computer after the photo has been taken, because, in the world of film photography, development and printing are subject only to the impersonal laws of chemistry and immune to human intervention. In fact, the way undeveloped film gets transformed into a negative, and the negative into a print, is anything but predetermined. One of the best known of Robert Frank's earlier photographs, *Men of Air*, shot in 1947, owes its otherworldly appearance to the fact that Frank mistakenly developed the film in a chemical bath set at the wrong temperature.

Most of the time, of course, the photographer's interventions are deliberate rather than accidental. At least as much artifice is exercised by the film photographer *after* the shutter has been released as has been exercised beforehand. Manipulating the chemistry of the film development process will alter the appearance of a negative. And then there are further decisions to be made in the darkroom that will affect the appearance of the print made from the negative—techniques such as "dodging" and "burning," choice of photographic paper, decisions about how to crop the print, etc. It was Ansel Adams, again, who said, "The negative is comparable to the composer's score and the print to its performance. Each performance differs in subtle ways."

In short, manipulation of reality is and always has been at the heart of photography. Film photographers manipulate, and digital photographers do, too. The discussion about which type of technology—film or digital—involves more or less manipulation is as uninteresting as it is unimportant. Indeed, one of the seldom-discussed ironies related to manipulating photographic images is that the purpose of manipulation—whether of a film-based or digital photograph—is often to make the finished product look *more*, not less, like what the photographer's eyes saw. Neither photographic film nor digital sensors can match the human eye's capacity to perceive dynamic range. Put simply, the eye is much better than a camera at handling scenes incorporating both very bright and very dark areas. Making such scenes appear more or less "realistic" in photographs requires manipulation both at the moment of taking the photograph and afterward. The image that appears on the cover of this book is a good example. My eye was able to simultaneously perceive the highlight details in the clouds and the shadow details in the dark pavement in the foreground. The digital sensor of my camera could not. Using exposure settings that created a faithful rendering of the clouds would produce a dark foreground lacking detail. On the other hand, settings that captured the detail in the foreground shadows would produce a washed out, overexposed sky. The only way to make the image appear reasonably life-like was to manipulate it with digital editing software.

The fundamental question faced by photographers isn't a technological one; and it certainly isn't about how to produce "accurate" copies of "reality." Rather, it's about what to photograph in the first place. Of all the photographer sees at any given moment, why would he photograph *this* as opposed to *that*? The question that so many people ask, today, about photographs—especially if they believe that the photograph was taken with a digital camera—namely, "Did it really look like that, or did you manipulate the photograph?" misses the point. (And in any case the question of what something "really looked like" is problematic, to say the least!) Better questions might be "What were you thinking when you took that photograph?" or "What did you want us to see in that photograph?" or "What were you trying to convey?" Judge the photograph on its power to convey an idea, a feeling or an emotion, and not on the kind of equipment used to take it or what was done with it in the darkroom or on a computer, or how "real" it is.

The work of contemporary Canadian photographer Jeff Wall addresses head-on the slipperiness of concepts like "representation" and "reality" in photographic art. His photograph *Mimic*, for example, is a mundane looking image of three people walking down a sidewalk in an urban, semi-industrial setting. It looks like a simple candid street shot—apparently capturing one fleeting moment of "real life" in the city. In fact the photograph was staged; it used actors and was rehearsed. To quote Adams one final time: "There are no rules for good photographs. There are only good photographs."

Moonrise, Saskatchewan, near Carlyle SK

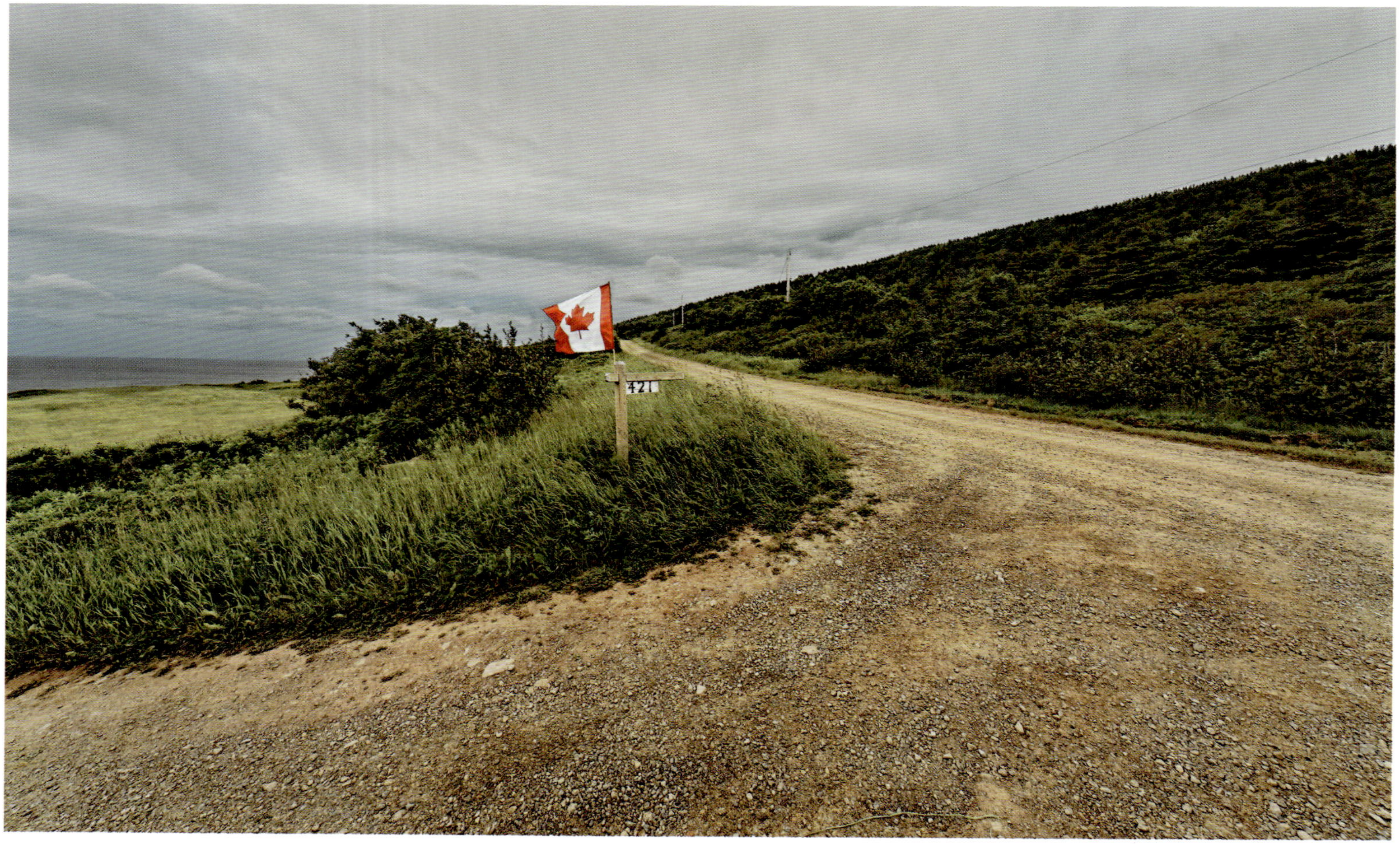

Flag, Broad Cove NS

The Land

WHEN GATHERING THE MATERIAL for *Roads* I divided the range of possible photographs into three categories: photographs where the road itself (and perhaps one or two structures beside the road) is the only obvious sign of human presence; photographs in rural areas, villages, and small towns, where you begin to see additional signs of human activity; and finally, photographs taken in larger towns and cities.

The photographs are presented in a sequence—"The Land," "Farm, Village, and Town," "Cities"—that reflects this way of organizing my thoughts, from images with the least intense signs of human intervention to those with the most.

In photographs of villages, towns, or cities one would expect to find roads. We're less likely to expect them in landscape photographs. When they do appear, they bring an element of incongruity, of surprise, and sometimes irony. That's why the visual impact of roads is often at its most dramatic in a landscape. The artificial lines and curves, the human engineering, stand out against the natural backdrop.

My eye is drawn to the geometry of converging perspective lines as a road fades off into the distance, as in *Oil Well* (p. 19), *Motorcyclist* (p. 68), *Lake Manitoba Narrows* (p.69) and *Eastbound* (p. 71). I like the rhythms that roads create through repetition, as in the guard-rail posts in *Percé Rock* (p. 61), and the corkscrew pattern of the highway in *Road to Point Lance* (p. 74). And there's nothing better than a road to provide a solid foundation for a photograph by sitting weightily at the bottom of the frame, with a kind of visual gravitas and the firm (almost literal) underpinning of the painted road lines. This you see in *Moonrise-Saskatchewan* (p. 33) and *Deadmoose Lake* (p. 56).

Because a road stands out in a landscape photograph, the viewer might wonder what the photographer had in mind. Why did he put a road in the middle of that composition? Or, as in the case of *Highway 430 near Barr'd Harbour* (p. 29), not only a road but also a telephone pole (the inescapable companion of the road).

The answer, in part, is purely in the visual terms described above. Roads work well as a compositional element in a landscape photograph. They give structure to an image—providing it with a human-made frame and guiding lines that lead the eye across or into the photograph. And the constructed or painted lines and curves—products of careful calculations by civil engineers—contrast nicely with the haphazard bumps, ridges, and undulations of the natural surroundings, as in *Willows* (p. 47).

The appeal of roads in a landscape is more than just visual, though. When I see a road in a landscape I see a story about human attempts to exploit and to "conquer" the landscape. A road, for better or for worse, is an enabler. It is often one of the first human marks on an otherwise untouched landscape. When we use the land—build on it, harvest from it, subdivide it, play in it, admire it, buy it, and sell it—the road is our way in, out, and through.

TransCanada Highway (p. 14)—shot from a ditch beside Ontario Highway 17 near the town of Terrace Bay—is a good example of the "conquest" story. One of the last sections of the TransCanada Highway to be completed—in the 1960s—was not far from this spot in northwestern Ontario, north of Lake Superior. Building a road through the rock outcroppings of the Precambrian Shield created challenges on a par with those presented by the Rocky Mountains.

The Rockies themselves provided—and continue to provide—their own challenges to development. About 2,400 kilometres west of Thunder Bay, the TransCanada Highway climbs to 1,300 metres above sea level when it crosses the Rockies through the Rogers Pass. In the late nineteenth and early twentieth centuries, scores of workers on the Canadian Pacific Railway line through the pass—a route now followed by the highway—were killed in avalanches. Today, Rogers Pass is home to the largest mobile avalanche control program in the world. The Canadian military uses 105mm howitzers to trigger avalanches under controlled conditions so that motorists are not caught in unexpected slides. As an additional precaution, avalanche sheds (*Avalanche Shed,* p. 44) are built out over the road in areas that are especially vulnerable to avalanches.

Rocks come tumbling down from the mountains, too, and rarely in a more spectacular manner than they did in 1903 in the coal-mining town of Frank, Alberta, near the eastern entrance to the Crowsnest Pass through the Rockies. At 4 am, while most of the town's six hundred residents slept, the top of Turtle Mountain—a slab of rock nearly a kilometre wide, 650 metres high, and 150 metres thick broke off and crashed into the valley of the Crowsnest River. It was the greatest landslide ever to occur in North America. Today the TransCanada Highway passes through the debris field—90 million tonnes of limestone that cover three square kilometres of valley floor (*Frank Slide*, p. 77; part of the slope of Turtle Mountain is visible on the right of the frame).

Roads are built not only to conquer the land but also to guard it *from* conquest. Such was the case of the Alaska Highway (*Alaska Highway—Otter Falls Cut-Off*, p. 76), which runs for over 2,000 kilometres from Dawson Creek, British Columbia, through the Yukon to Delta Junction in Alaska. When construction of the road was first proposed in the 1920s the Canadian government showed little interest, seeing no point in a major construction project that would benefit just a few thousand Yukon residents. But the attack by Japan on Pearl Harbor during the Second World War, coupled with fears that the Japanese might threaten the west coast of North America, caused a change of views. The US Army offered to assume the entire cost of building the highway, and Canada agreed to allow construction through its territory. Work began early in 1942 and was completed just nine months later. Otter Falls, the destination of the side-road referred to in the photograph, acquired some fame in the 1950s when a new Canadian five-dollar bill featured a view of the falls on the back.

Farther north, the Dempster Highway (*Dempster Highway*, p. 45 ; *Mackenzie River*, p. 72) was a response to early efforts to find oil and gas in the Mackenzie River Delta in the 1950s. The road runs for 730 kilometres from Dawson City up into the Arctic, terminating at Inuvik, Northwest Territories, on the east side of the Mackenzie Delta. The road sits atop an unusually thick layer of gravel—more than two metres deep in some places—to insulate the permafrost, which is all that keeps the Dempster Highway from sinking into the ground. Much of the highway follows an old dog-sled trail that was used by its namesake, Inspector William Dempster of the RCMP. The Canadian government first announced its intention to build the road in 1958, but it wasn't officially opened until 1979, when it became Canada's first all-weather highway to cross the Arctic Circle.

The Confederation Bridge (*Confederation Bridge*, p. 89) was even longer in the making. The idea of a fixed link across the Strait of Northumberland between Prince Edward Island and the mainland was first discussed in the 1870s. Construction of the thirteen-kilometre bridge began more than a century later in 1993, and the bridge opened in 1997. When Prince Edward Island agreed to join Canada as its seventh province in 1873, it did so on the basis of a pledge by the Canadian government to "assume and defray all the charges for [...] Efficient Steam Service for the conveyance of mails and passengers" between the island and the mainland. The Canadian government provided this service by ferry until the opening of the bridge. Before construction was launched, the Canadian government had to seek an amendment to the Canadian Constitution to ensure that the existence of the bridge would relieve it of responsibility to provide a maritime link. A Constitutional Amendment was enacted in 1993 stating that "a fixed crossing joining the Island to the mainland may be substituted for … steam service."

Roads both determine and follow patterns of human industry and settlement. Over the short term the extent of the road network limits where we choose to live, and the places where we can exploit the earth's resources. But as our numbers and our appetites grow, we push outward, building new roads to land that was once unreachable. And so roads form a tracery of human ambition (or greed, some might say). They mark our opposing desires to be both on the move and to be settled.

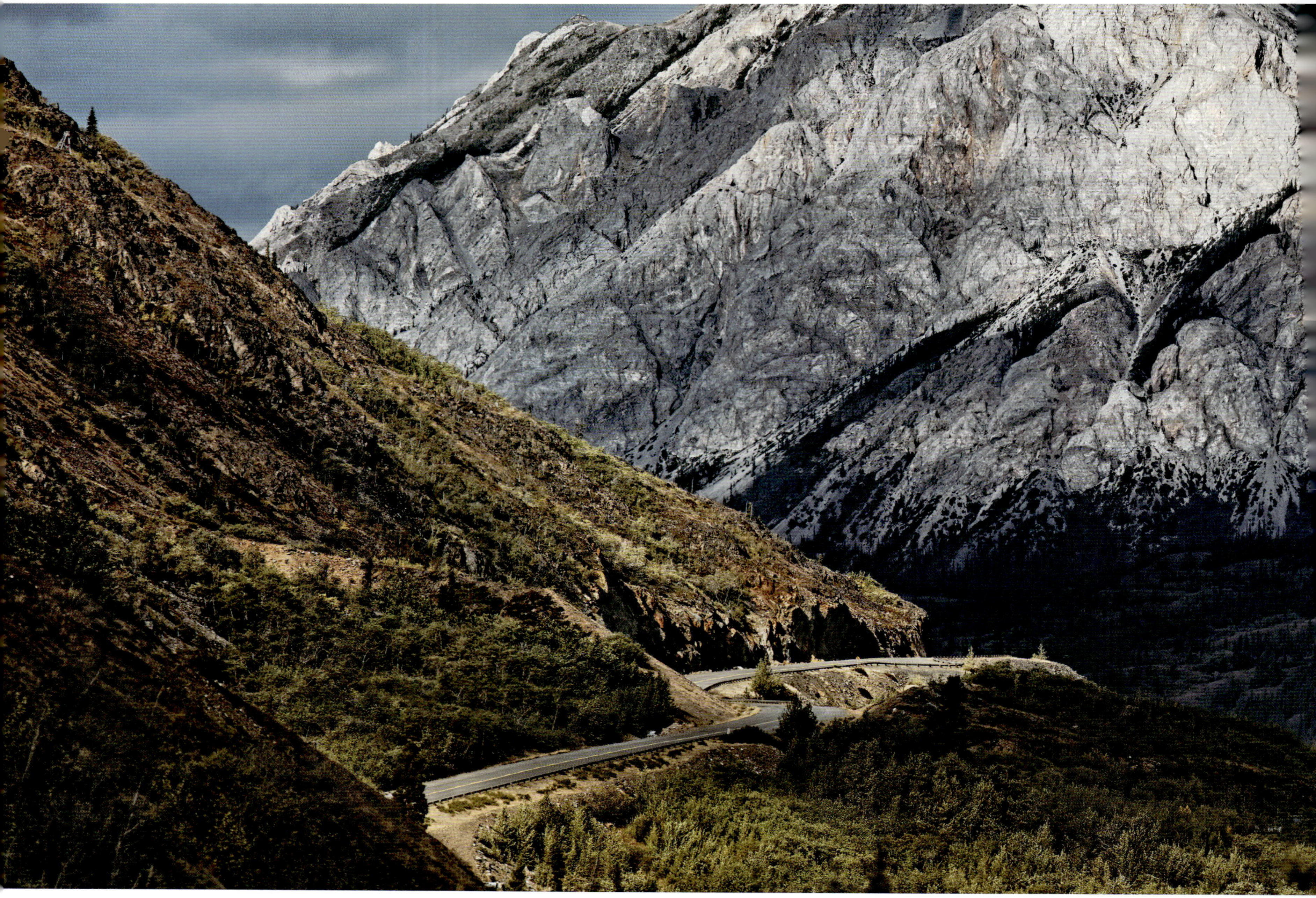

Road from Skagway, Looking North, Yukon Highway 2, near British Columbia Border YK

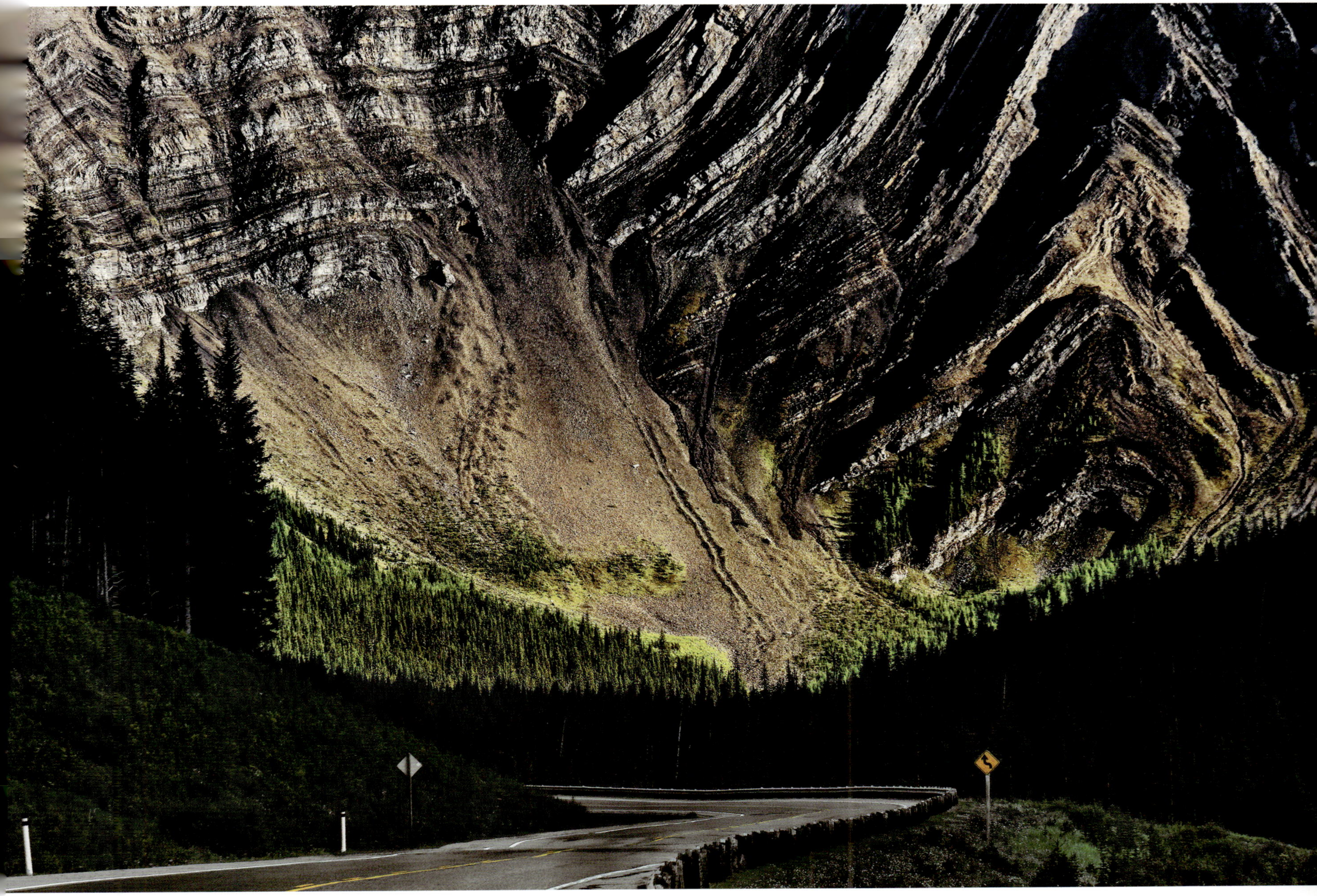

Kananaskis Valley, Alberta Route 40, Kananaskis Valley AB

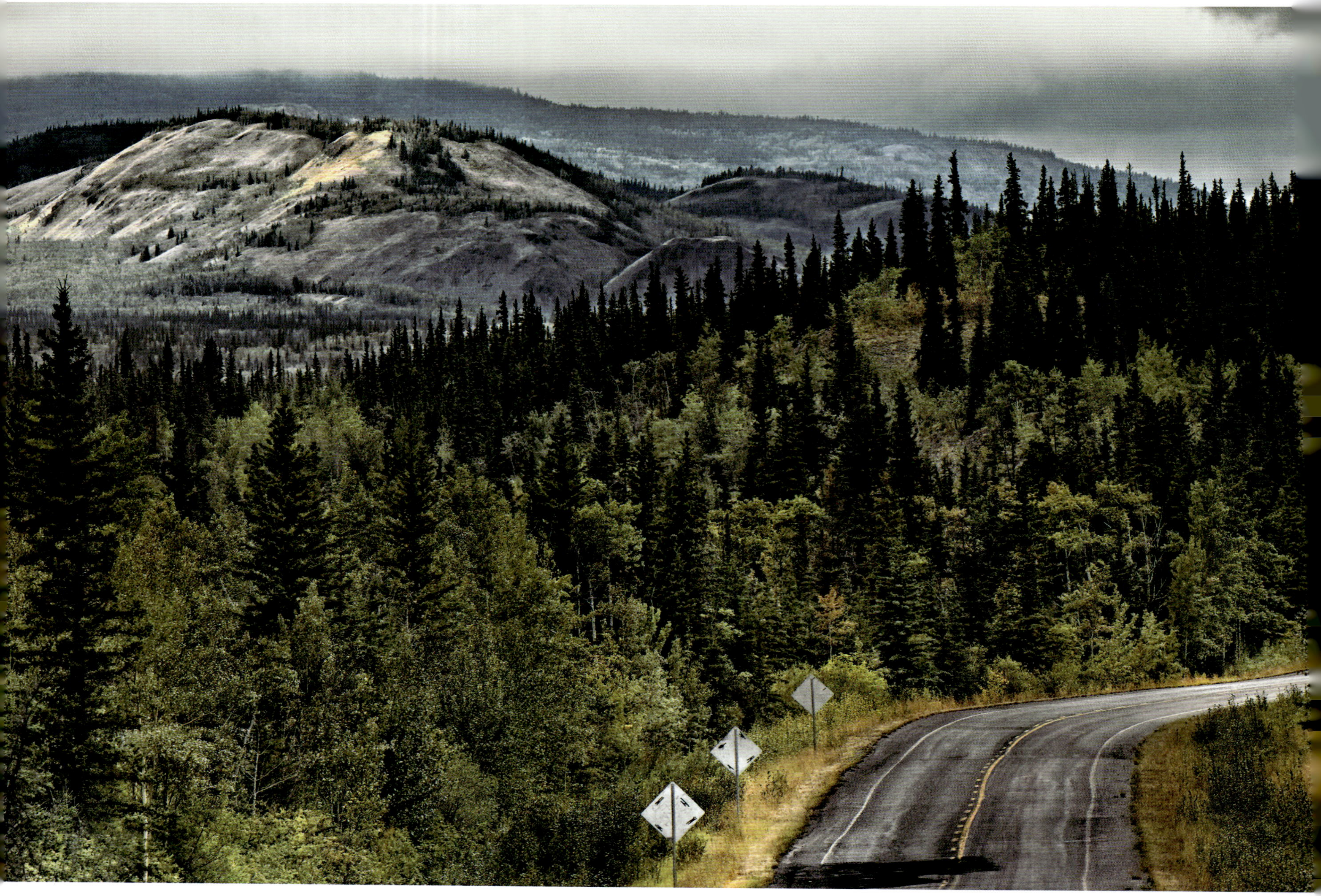

Three Road Signs, near Carmacks YK

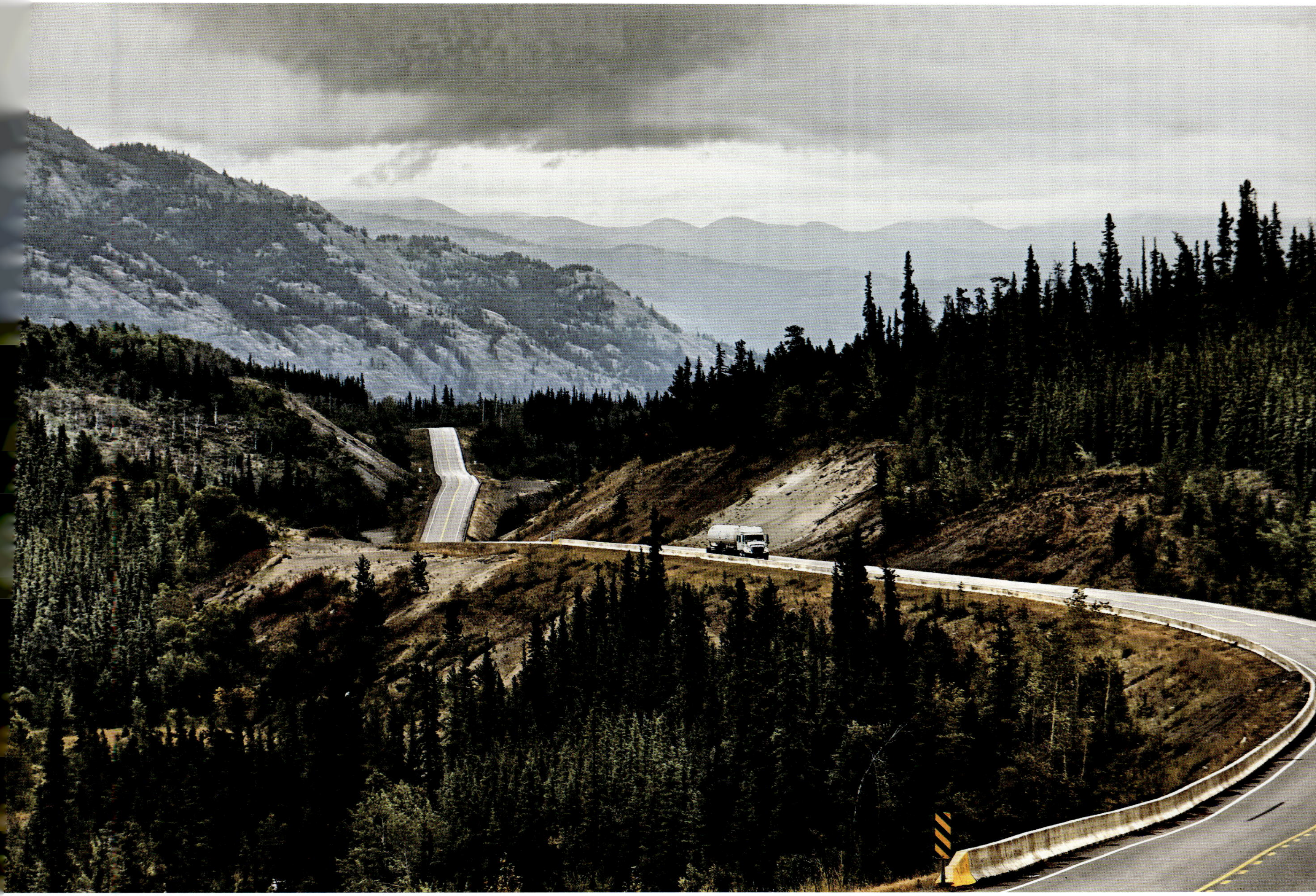

Tanker Truck, Alaska Highway, south of Whitehorse YK

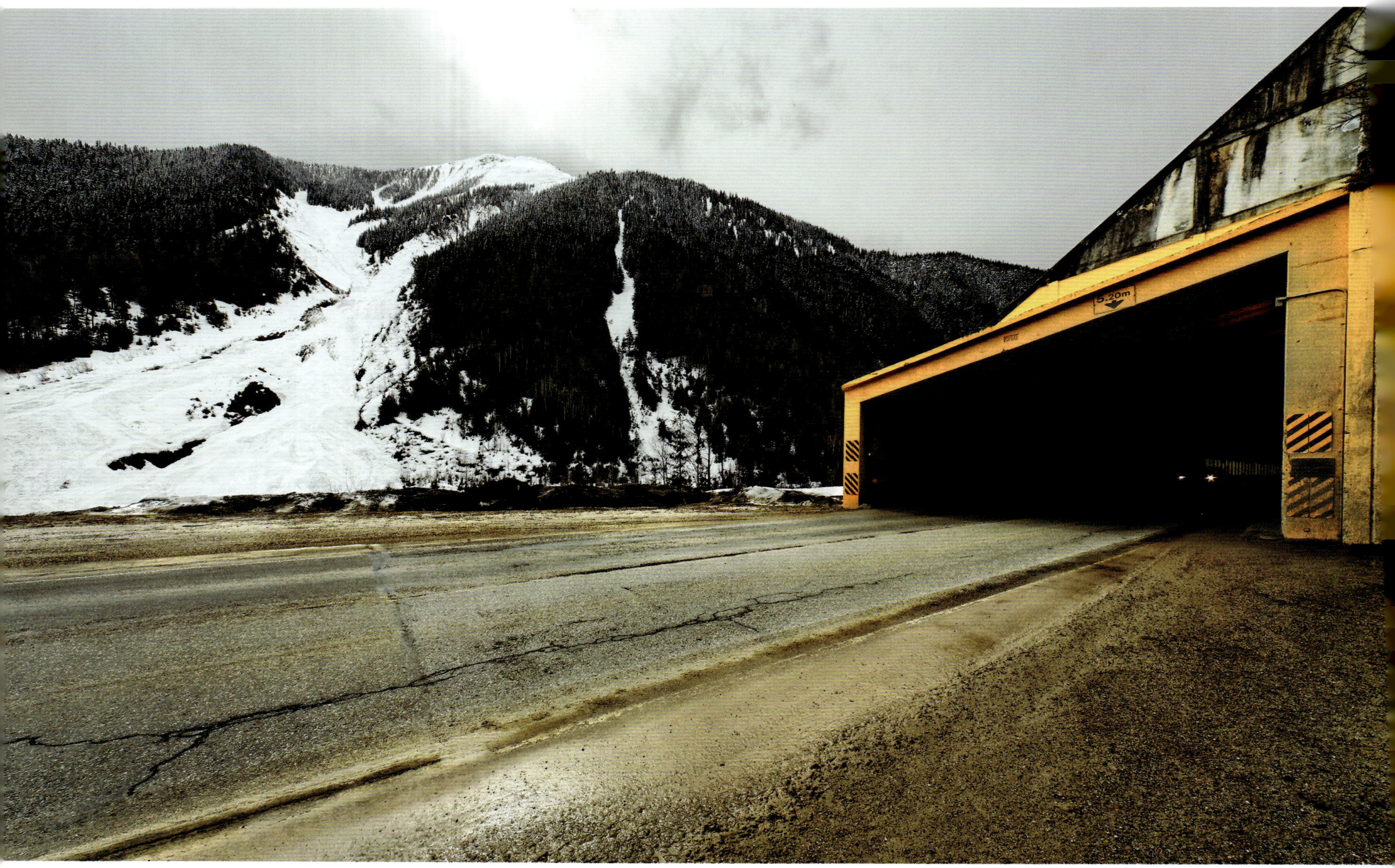

Avalanche Shed, near Canyon Hot Springs BC

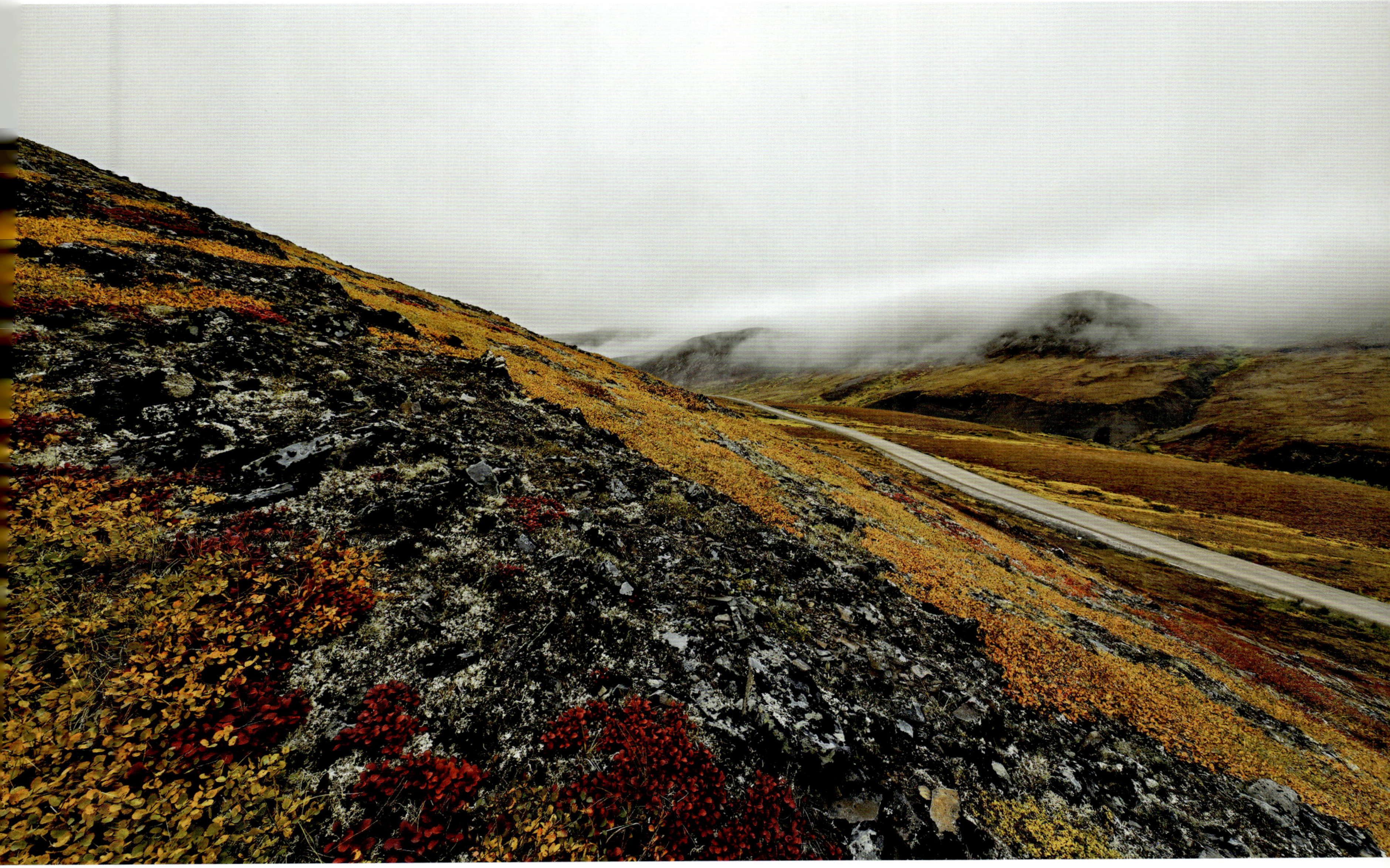

Dempster Highway, near the Yukon border NT

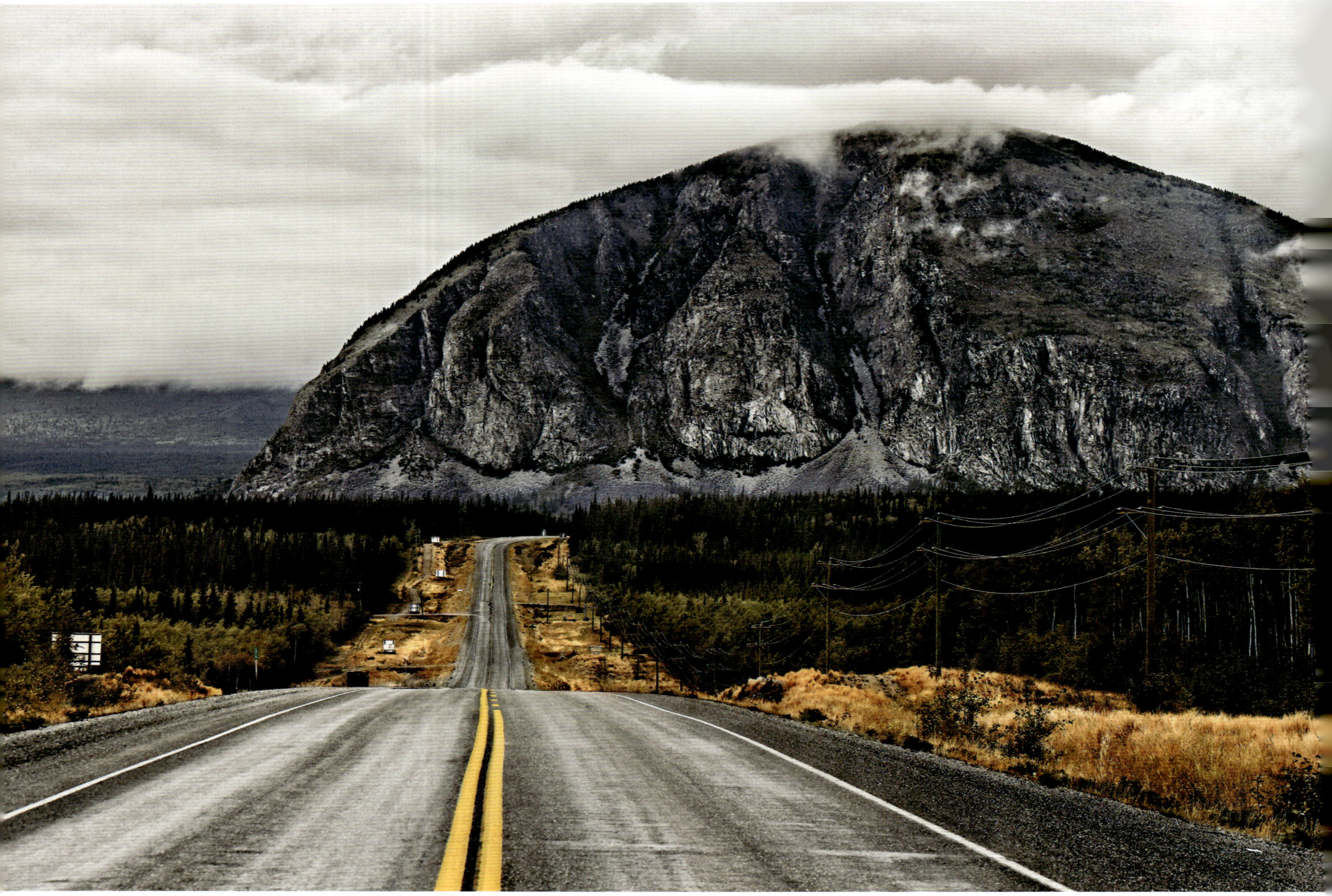

Mount Bark, Haines Junction YK

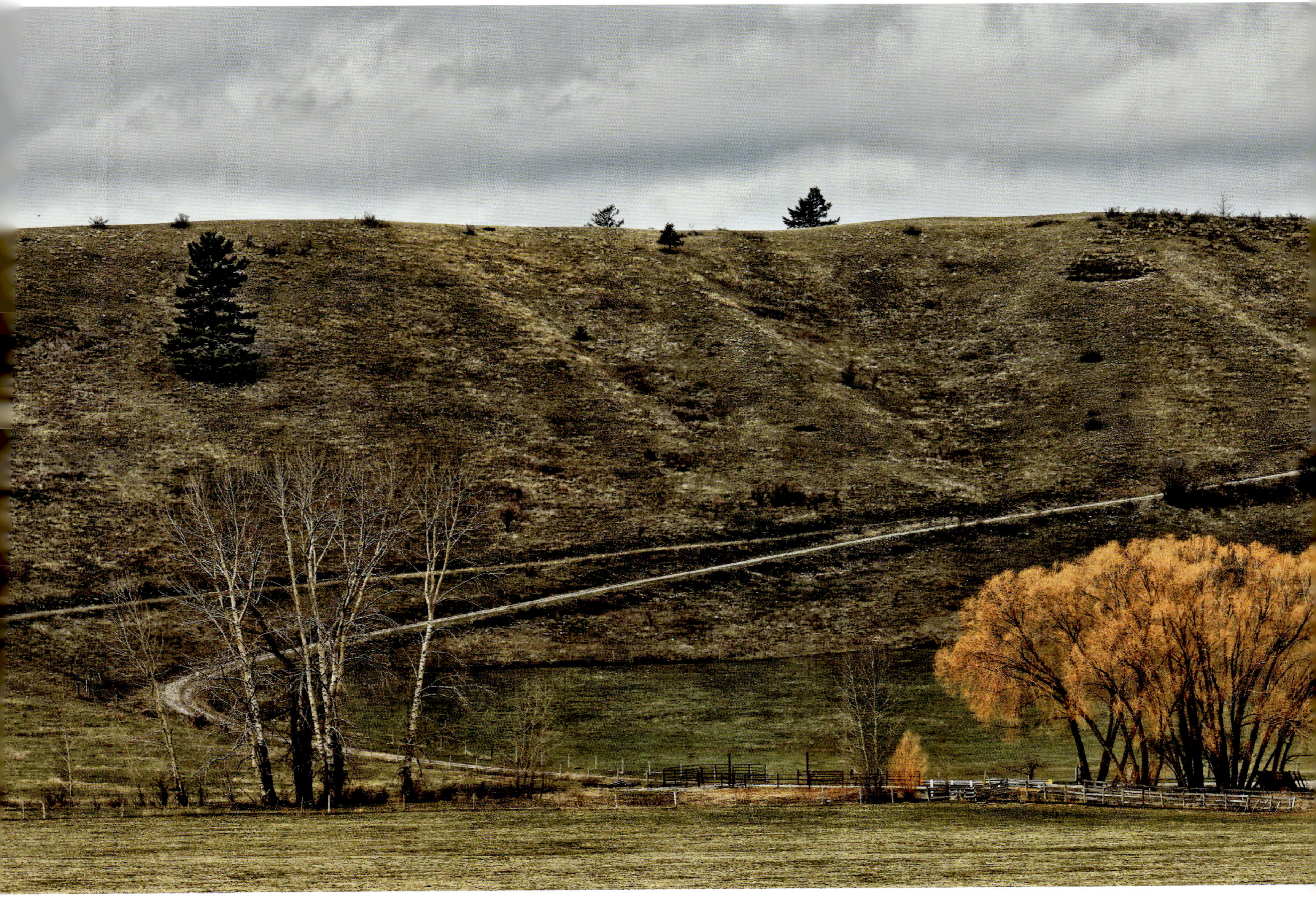

Willows, Roosville BC

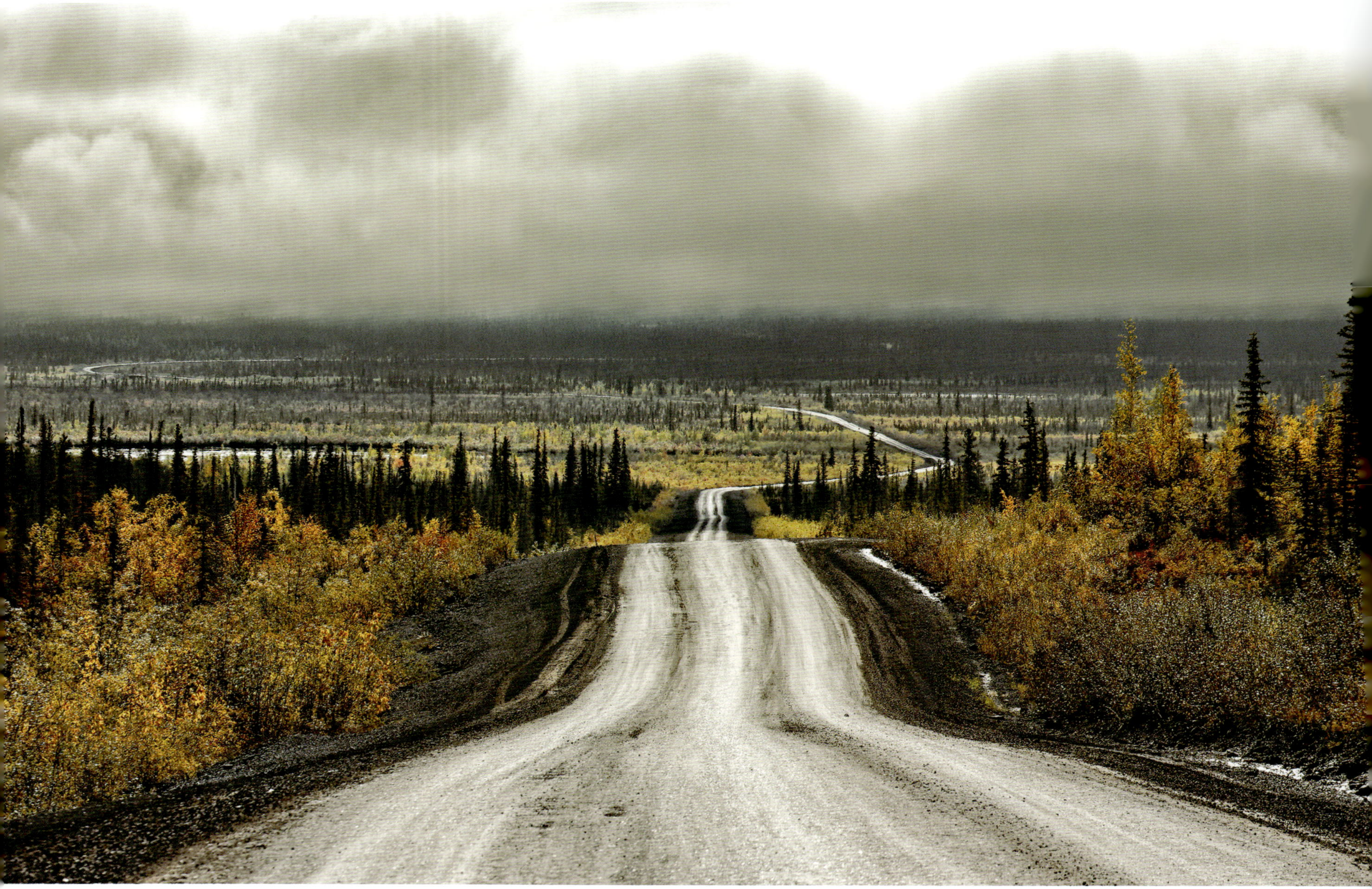

Dempster Highway, near Inuvik NT

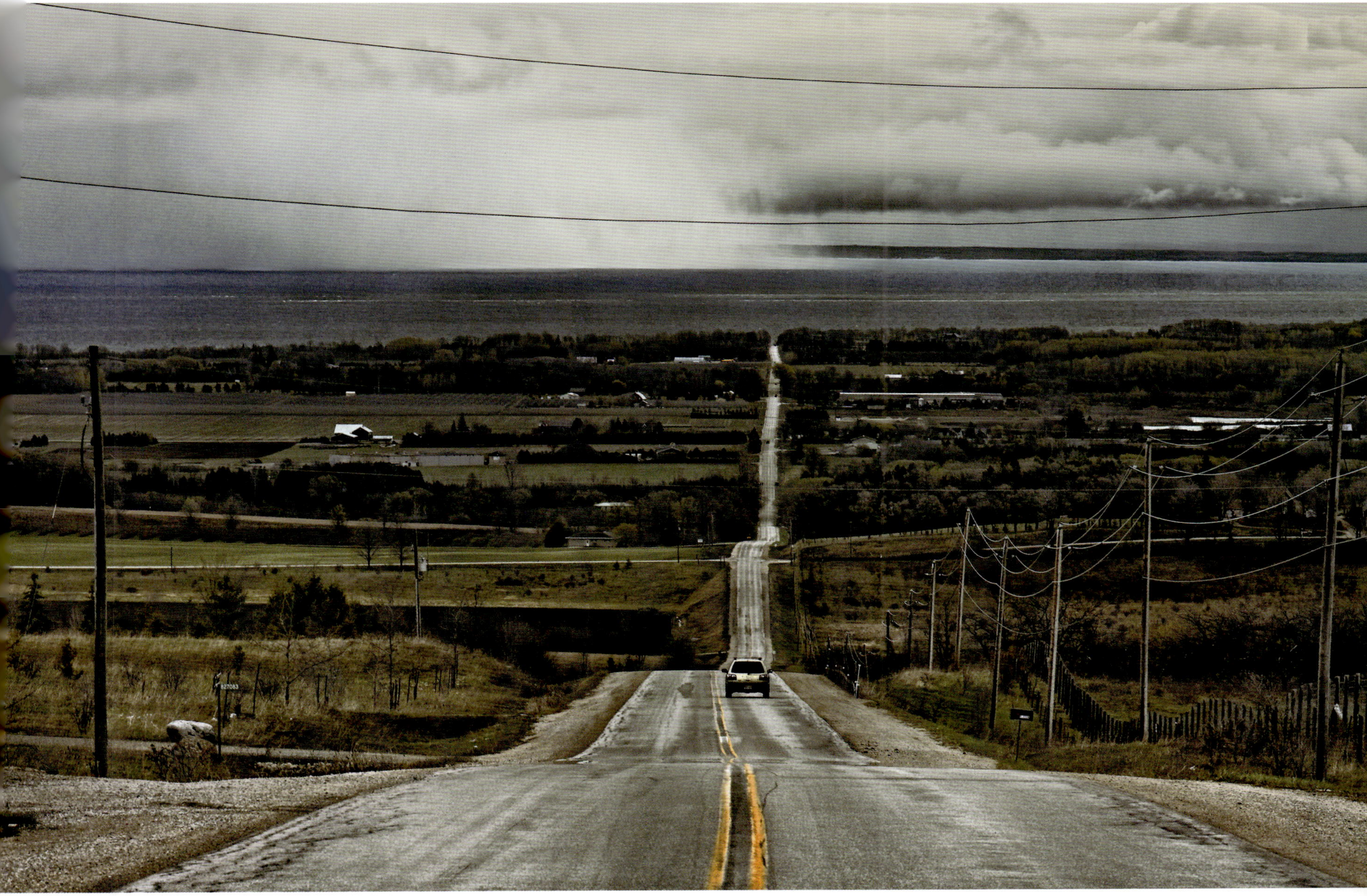

Squall, County Road 40 near Fairmount ON

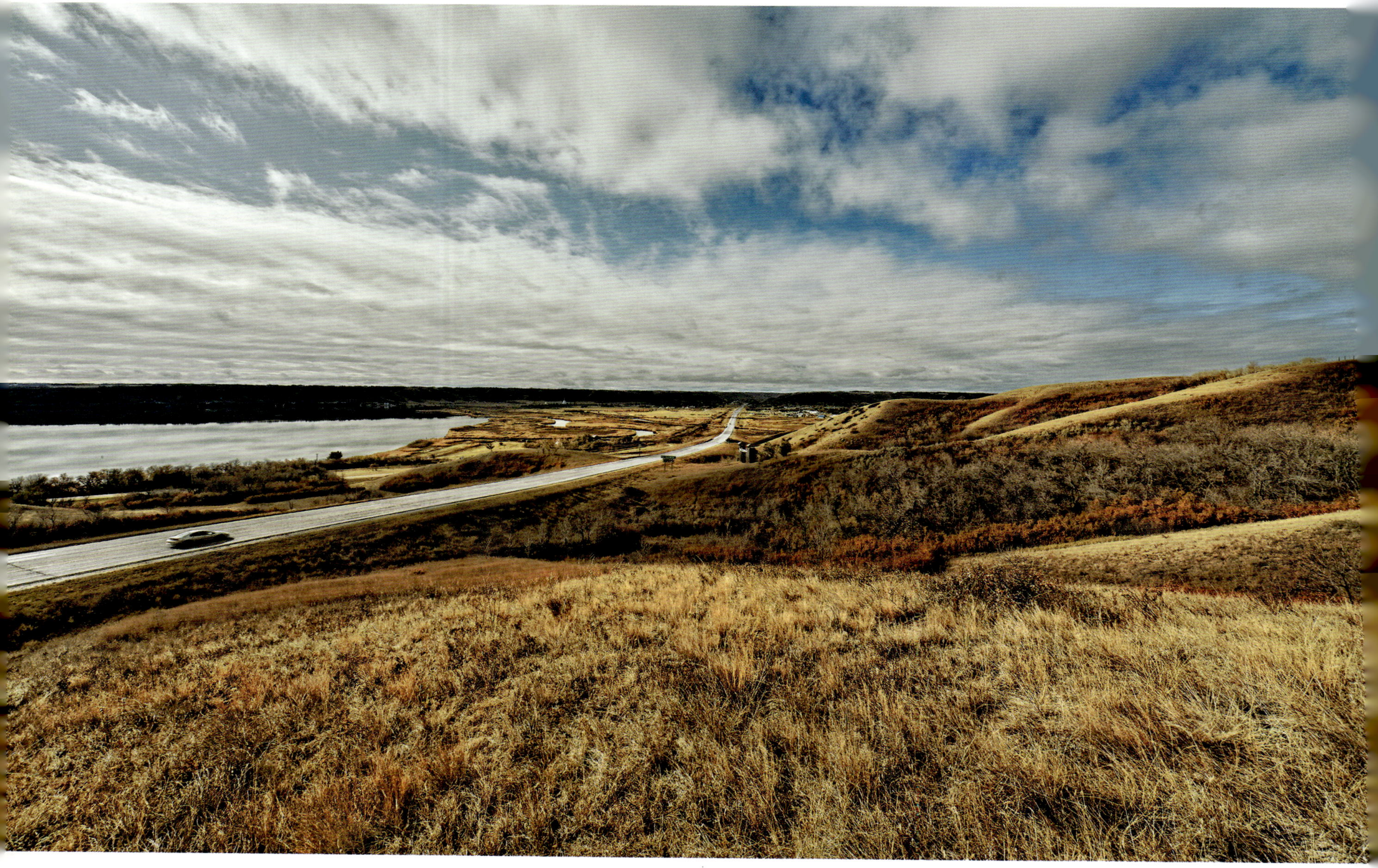

Qu'Appelle Valley, Fort Qu'Appelle SK

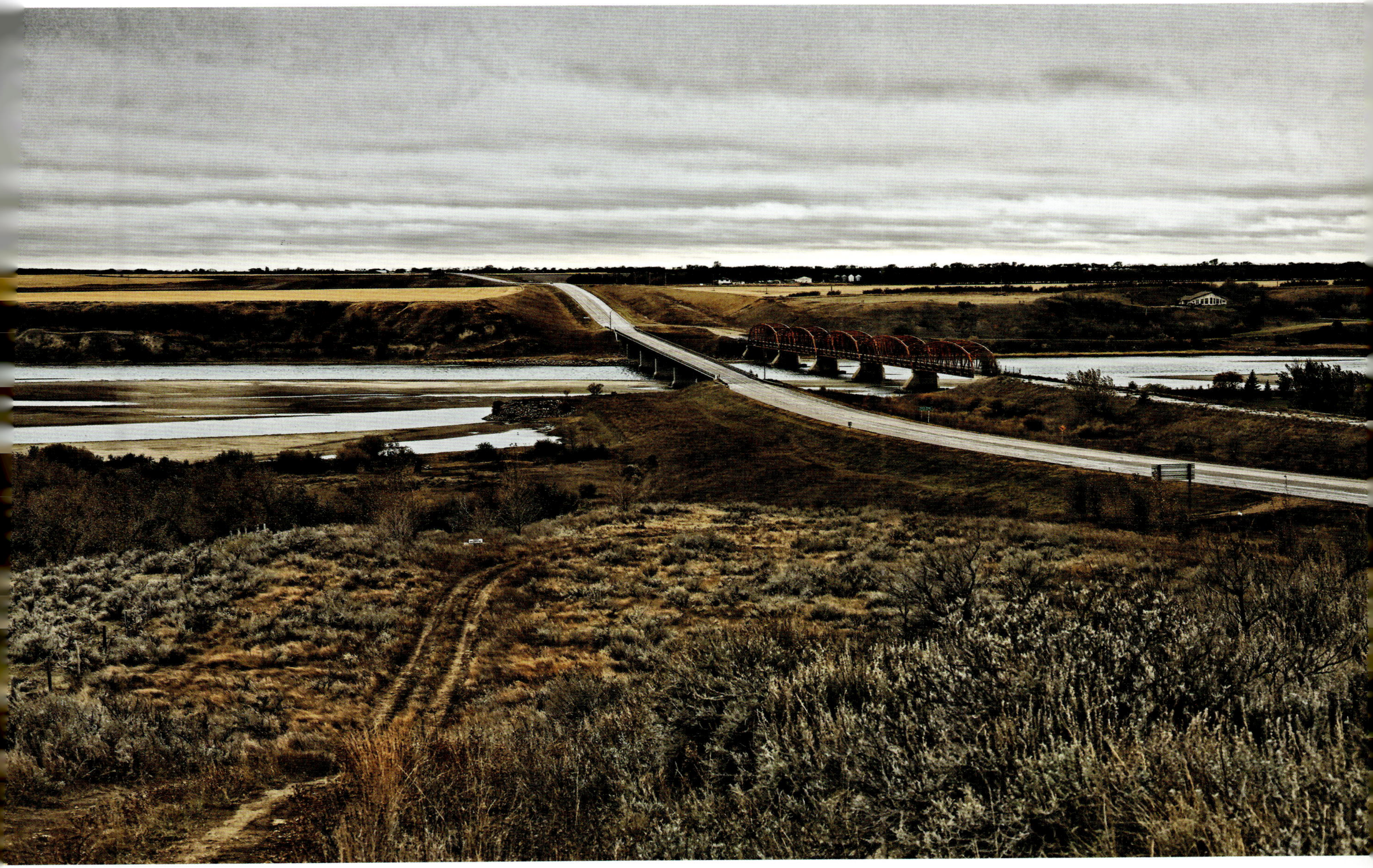

Bridges, South Saskatchewan River, near Outlook SK

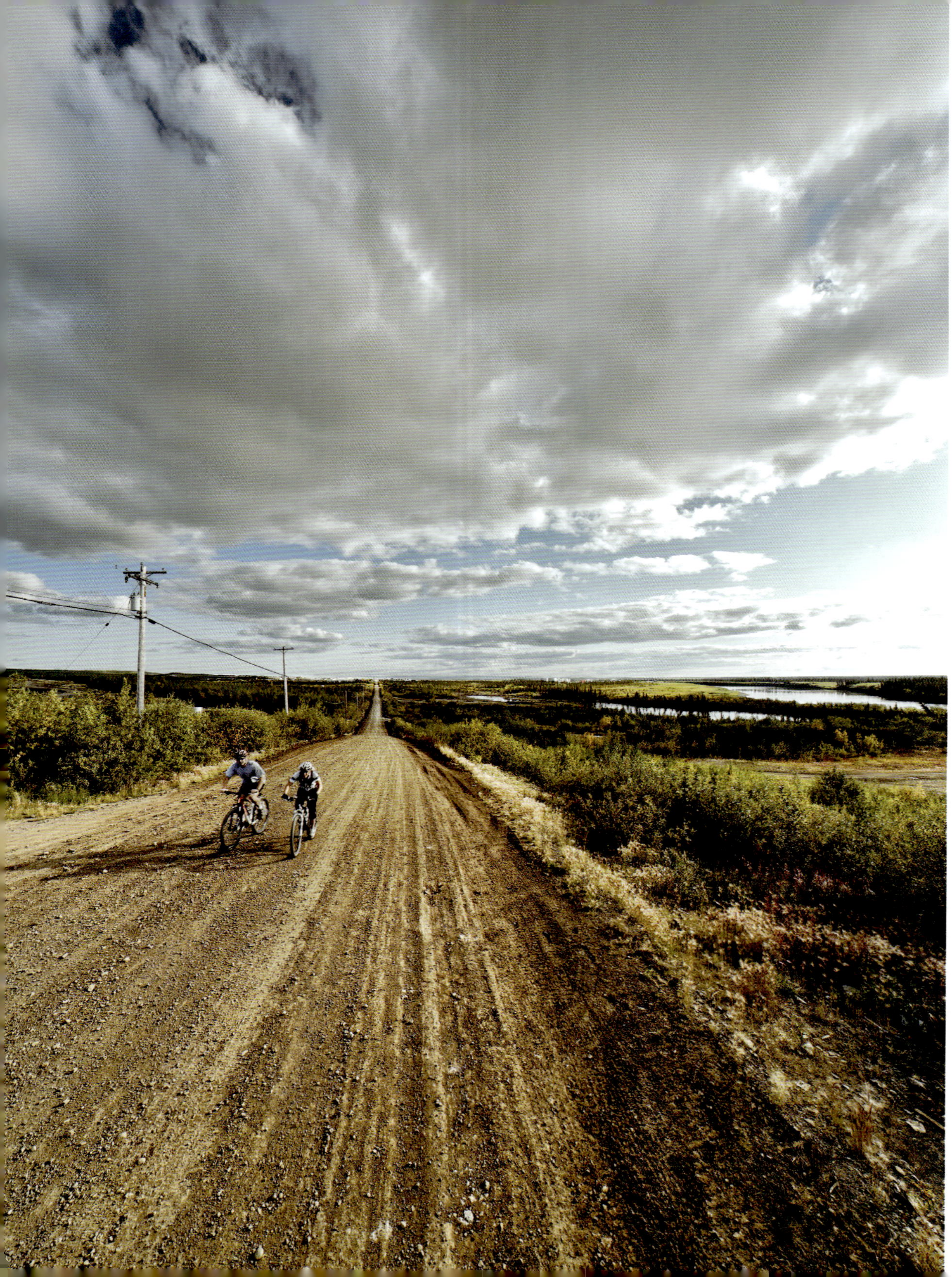

Cyclists, Inuvik NT

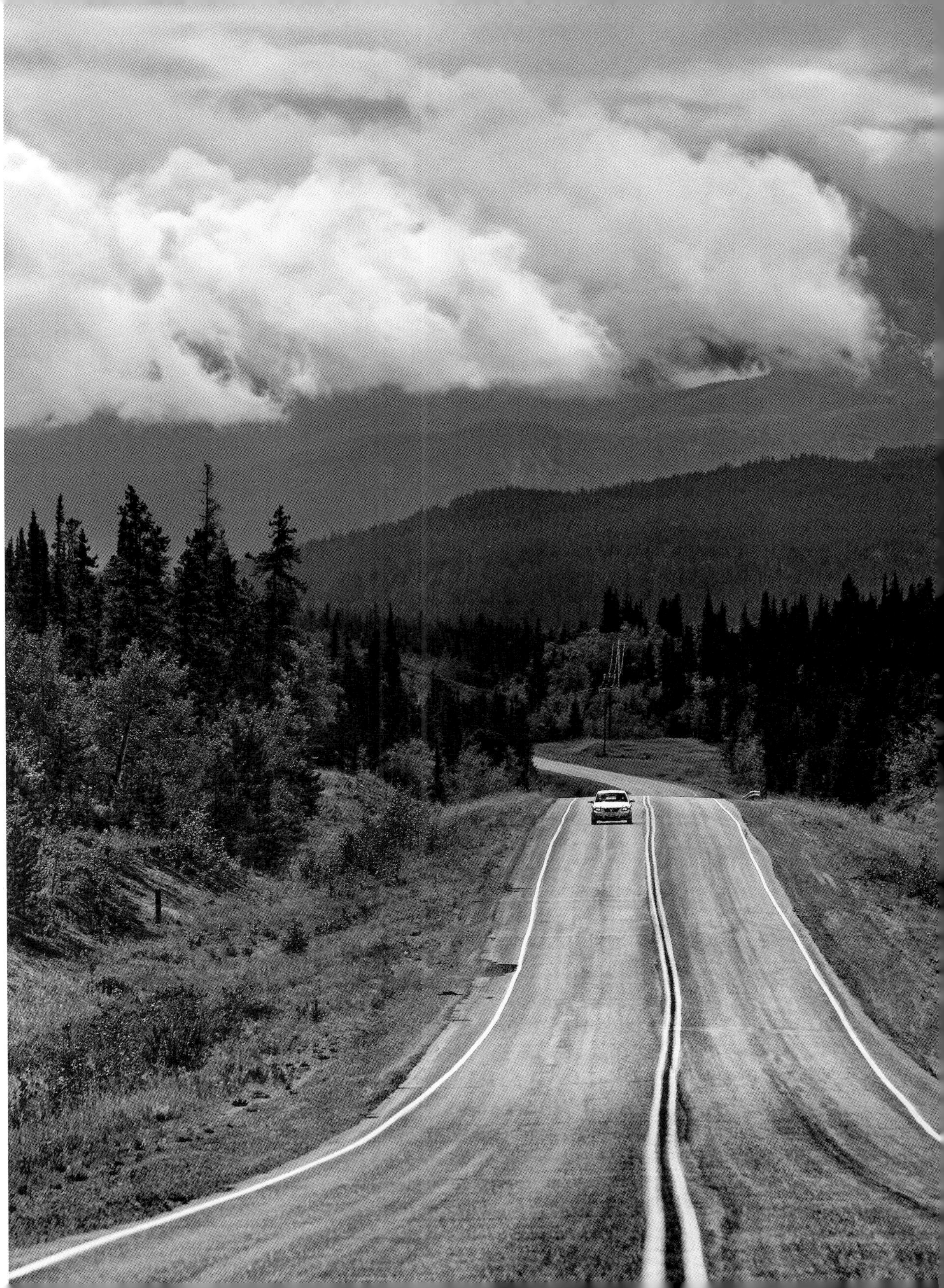

Double Line, near Robinson YK

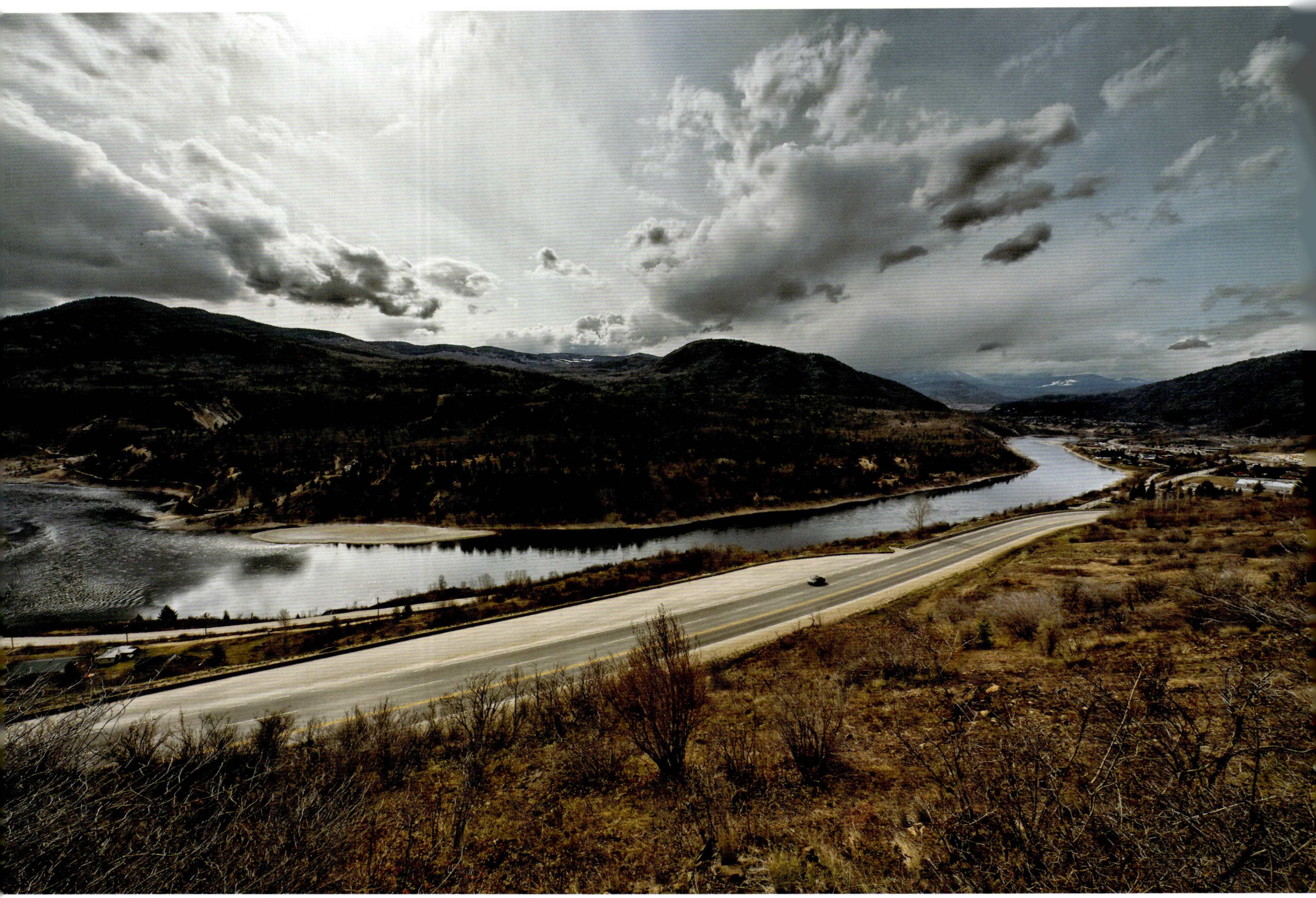

Columbia River, near Montrose BC

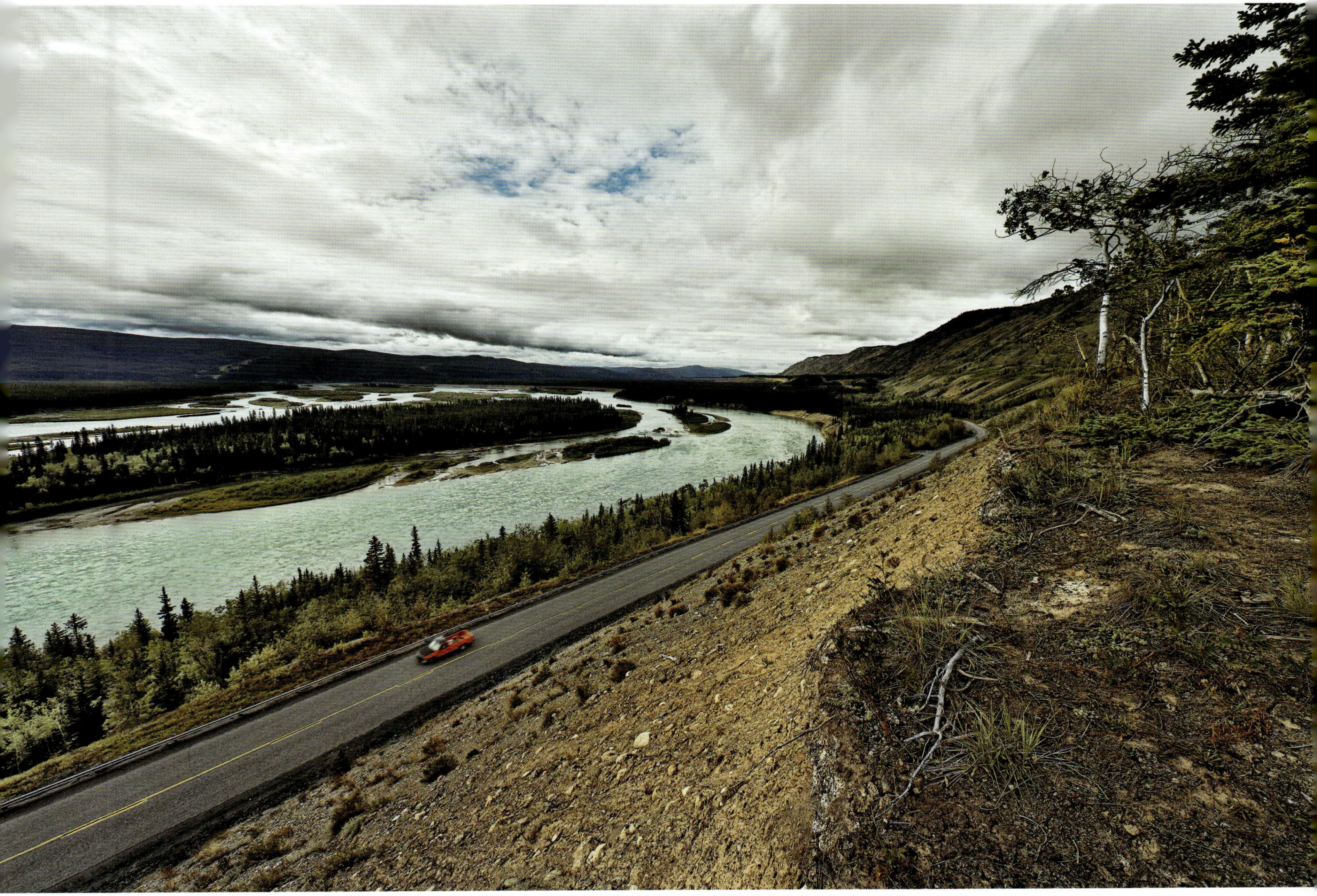

Yukon River, Klondike Highway north of Five Finger Rapids YK

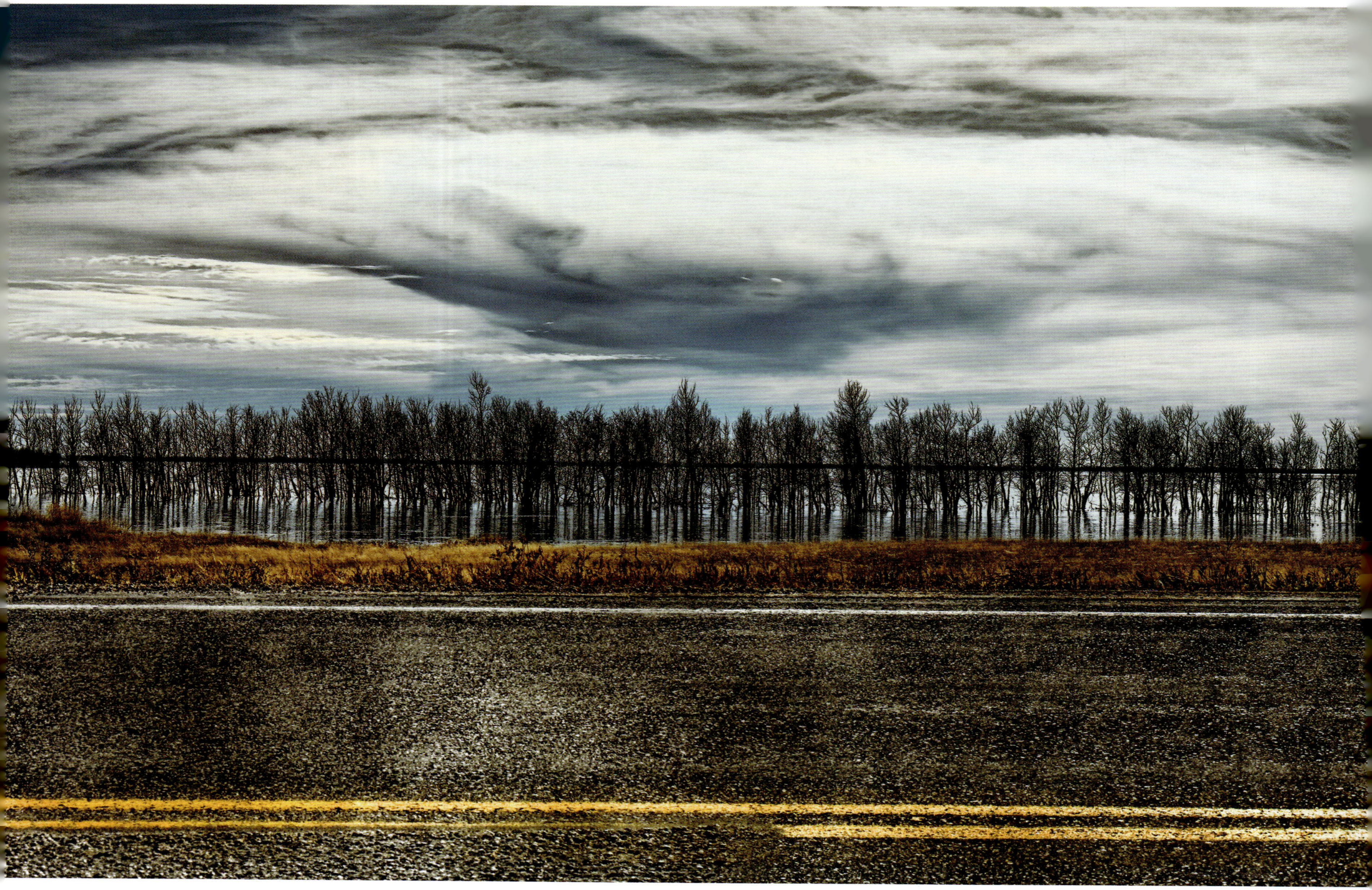

Deadmoose Lake, Marysburg SK

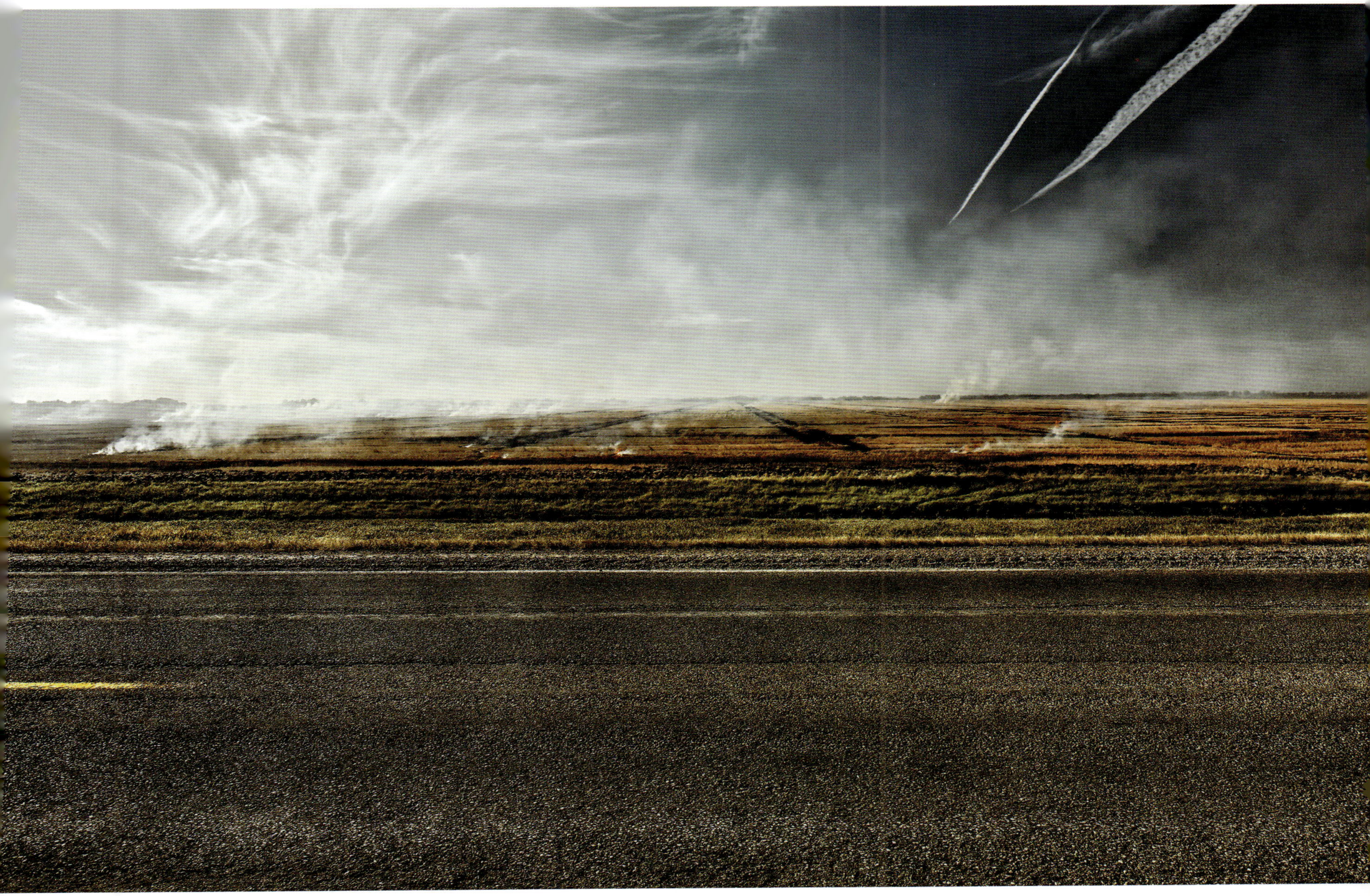

Fields on Fire, near Morris MB

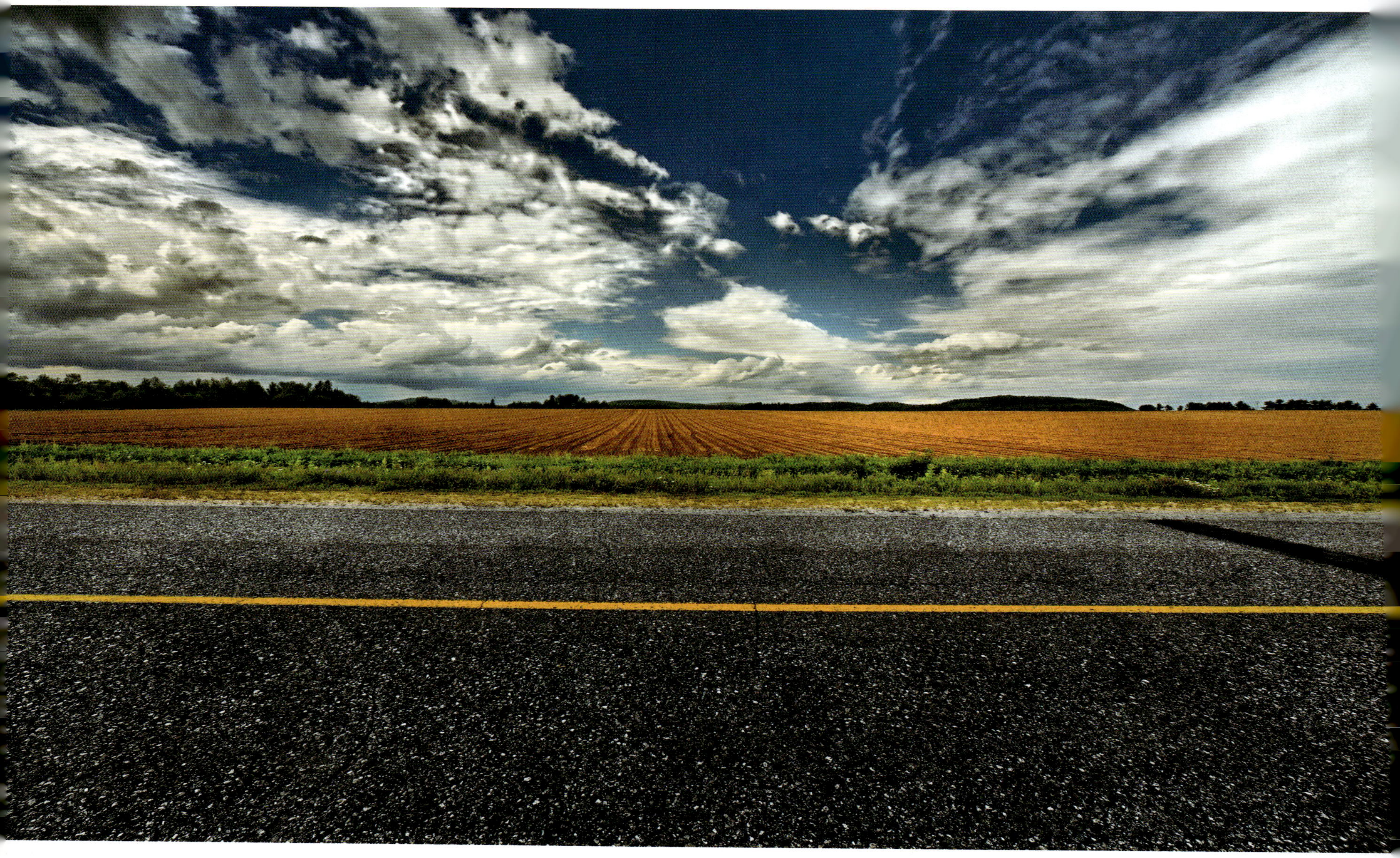

Potato Field, near Notre Dame de la Paix QC

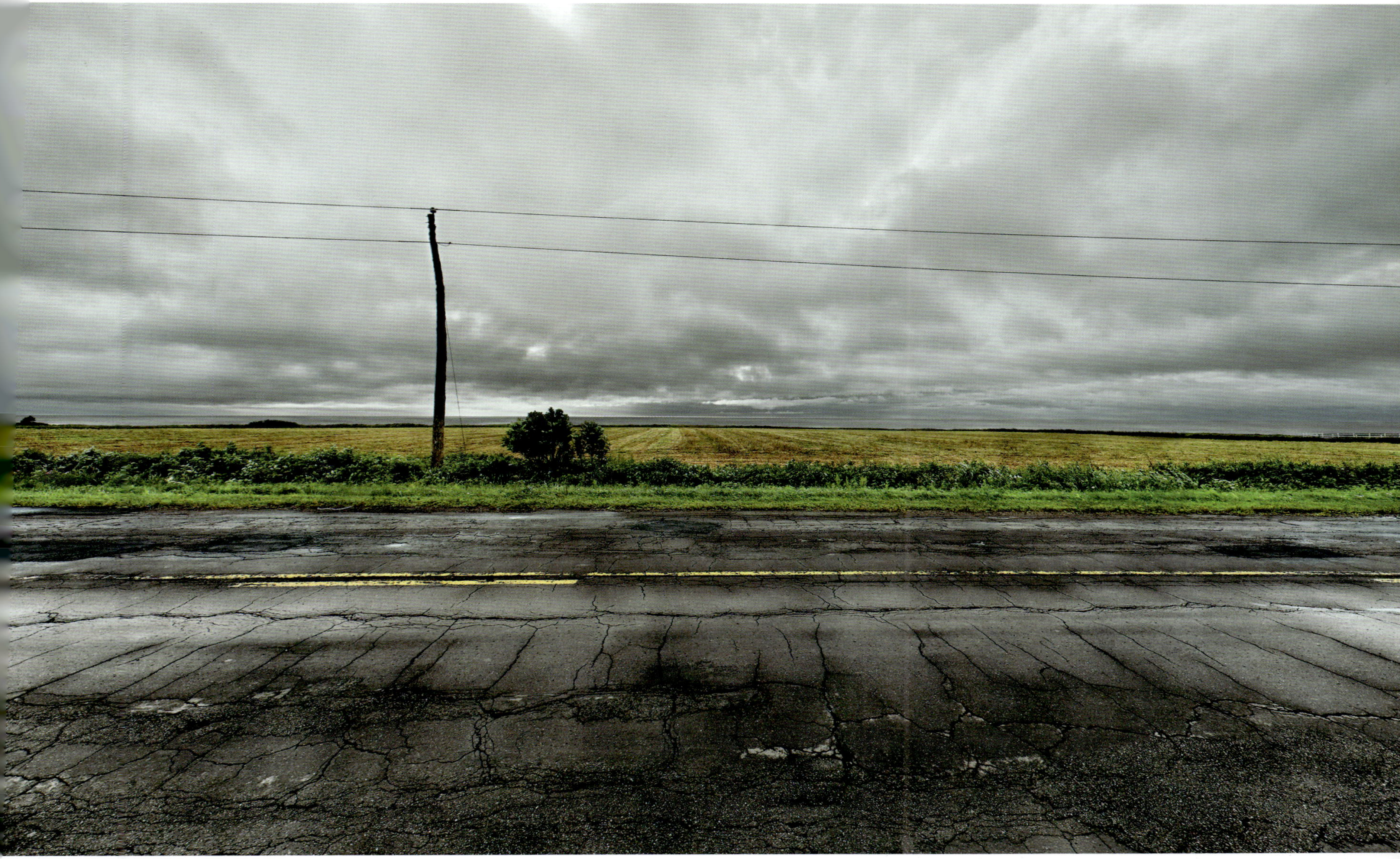

West Cape, West Cape PEI

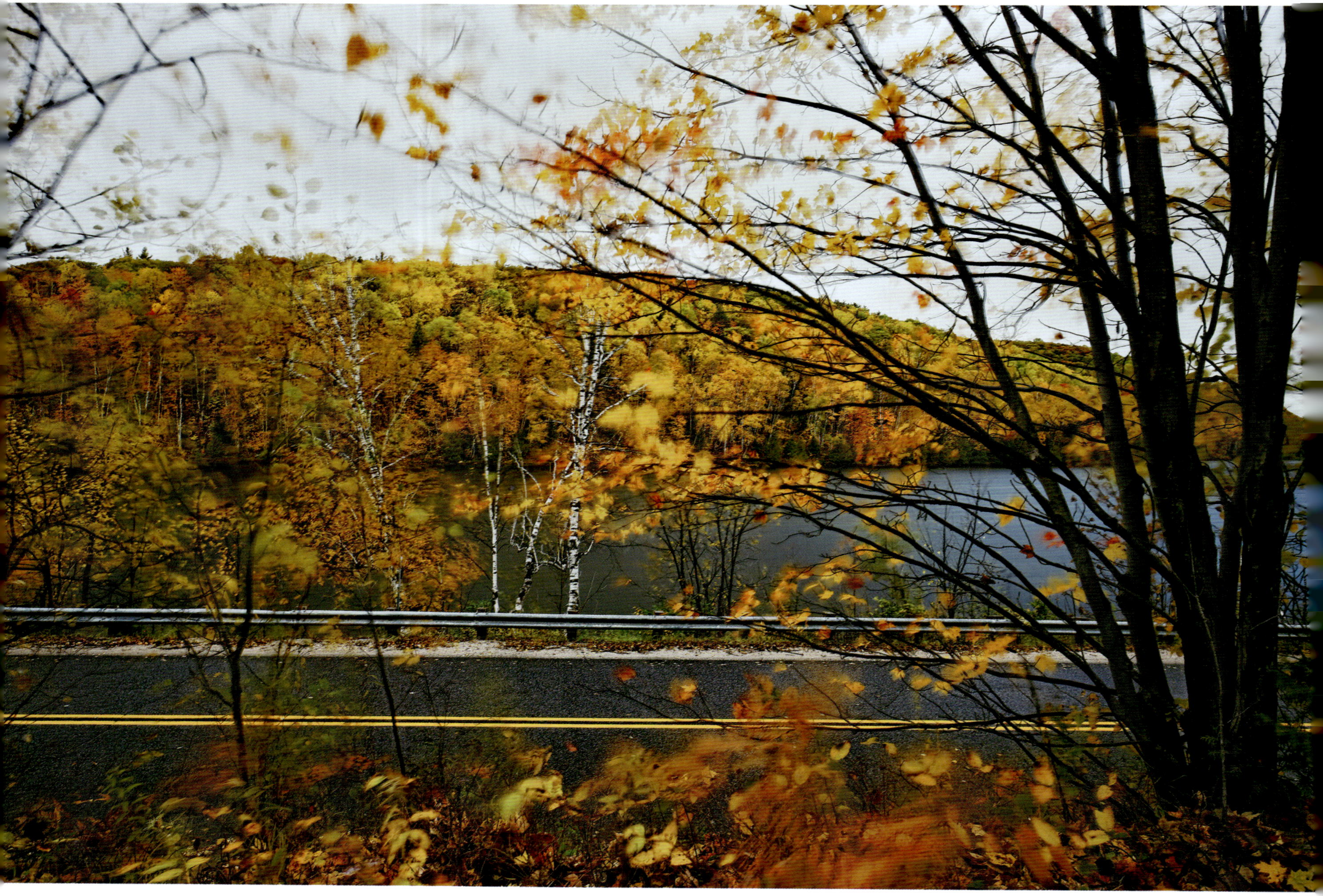

Lola Lake, East Aldfield QC

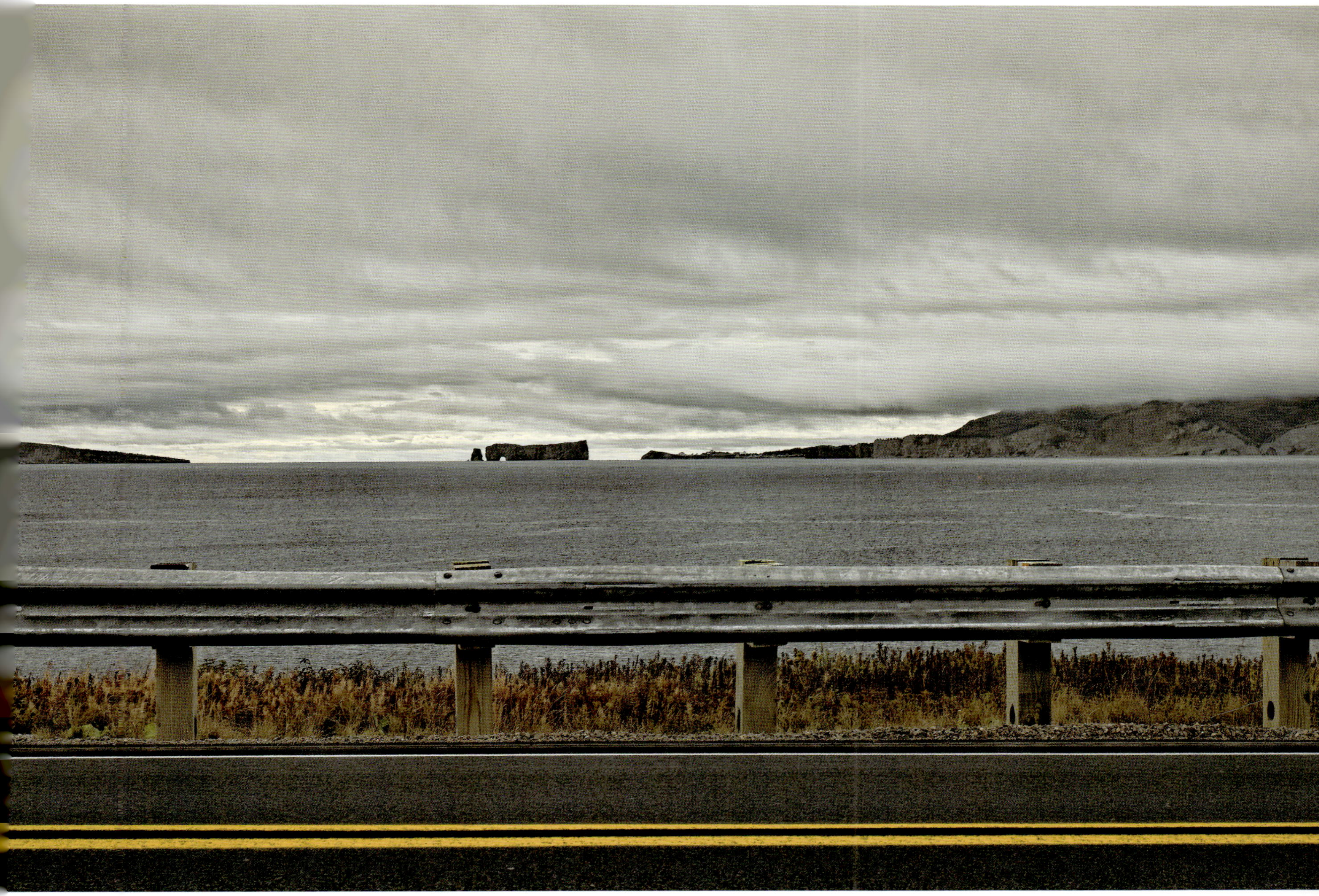

Percé Rock, Pointe-Saint-Pierre QC

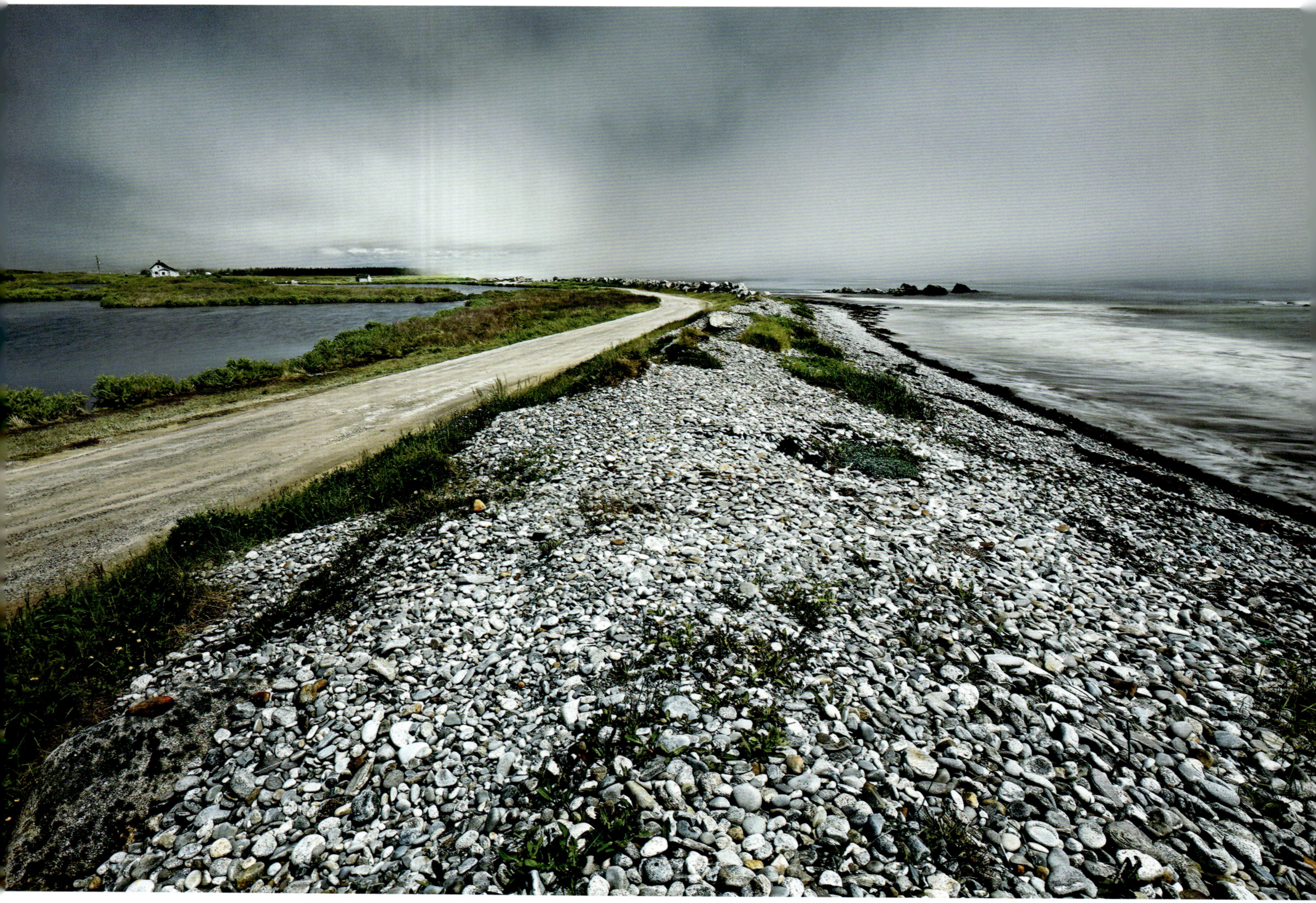

Hawk Beach, Hawk Beach NS

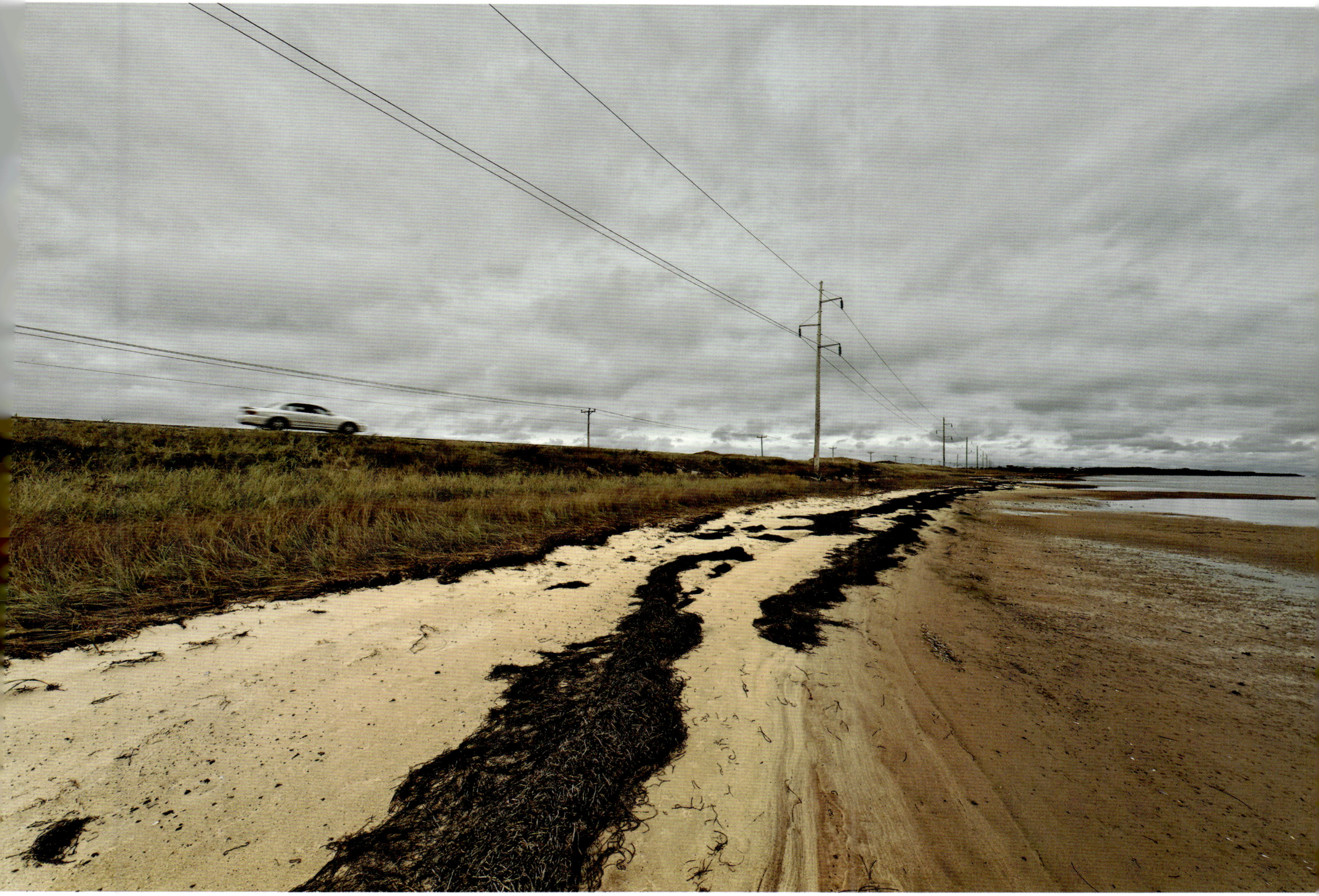

Approaching Pointe aux Loups, Magdalen Islands QC

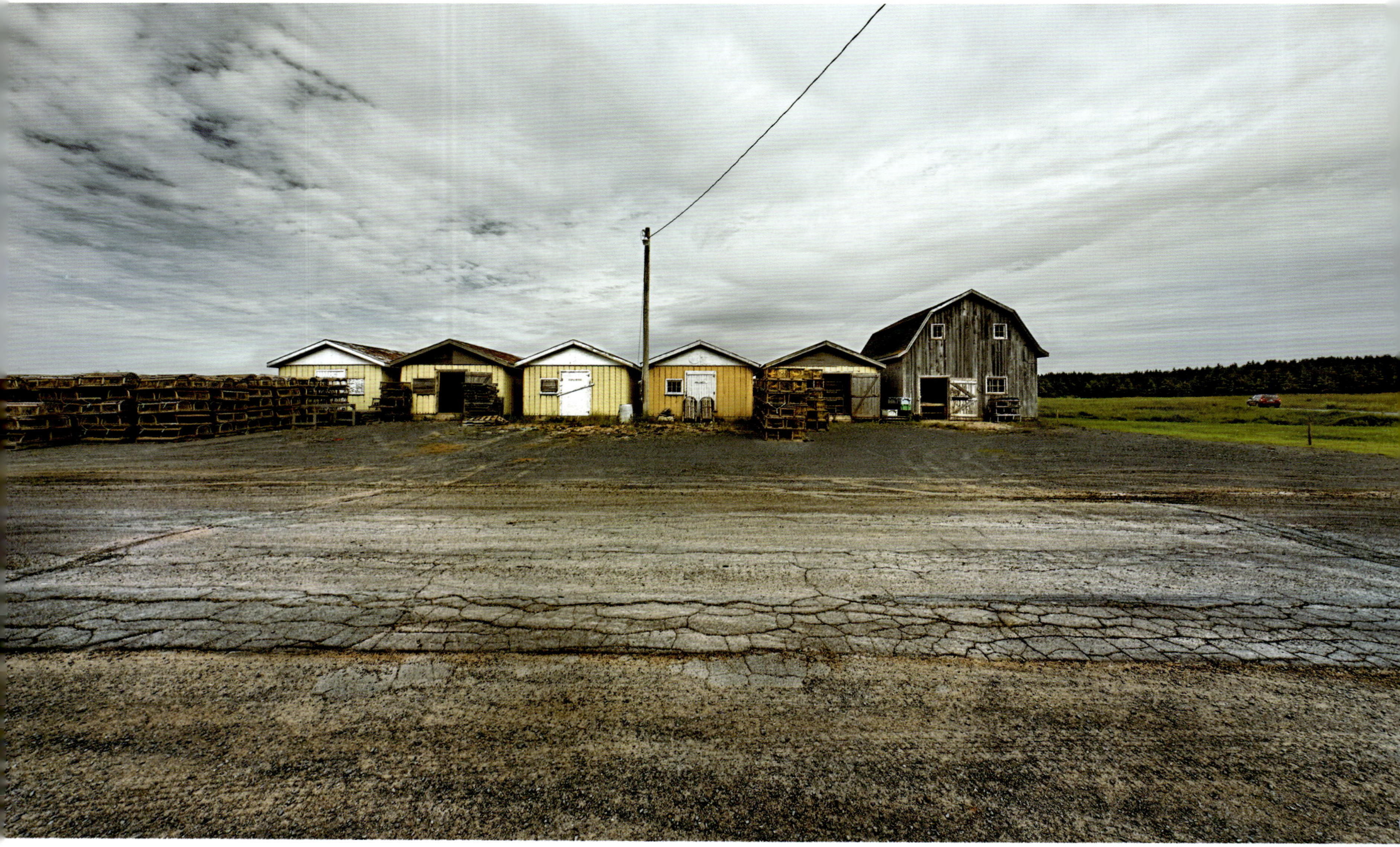

Sheds and Lobster Traps, Naufrage PEI

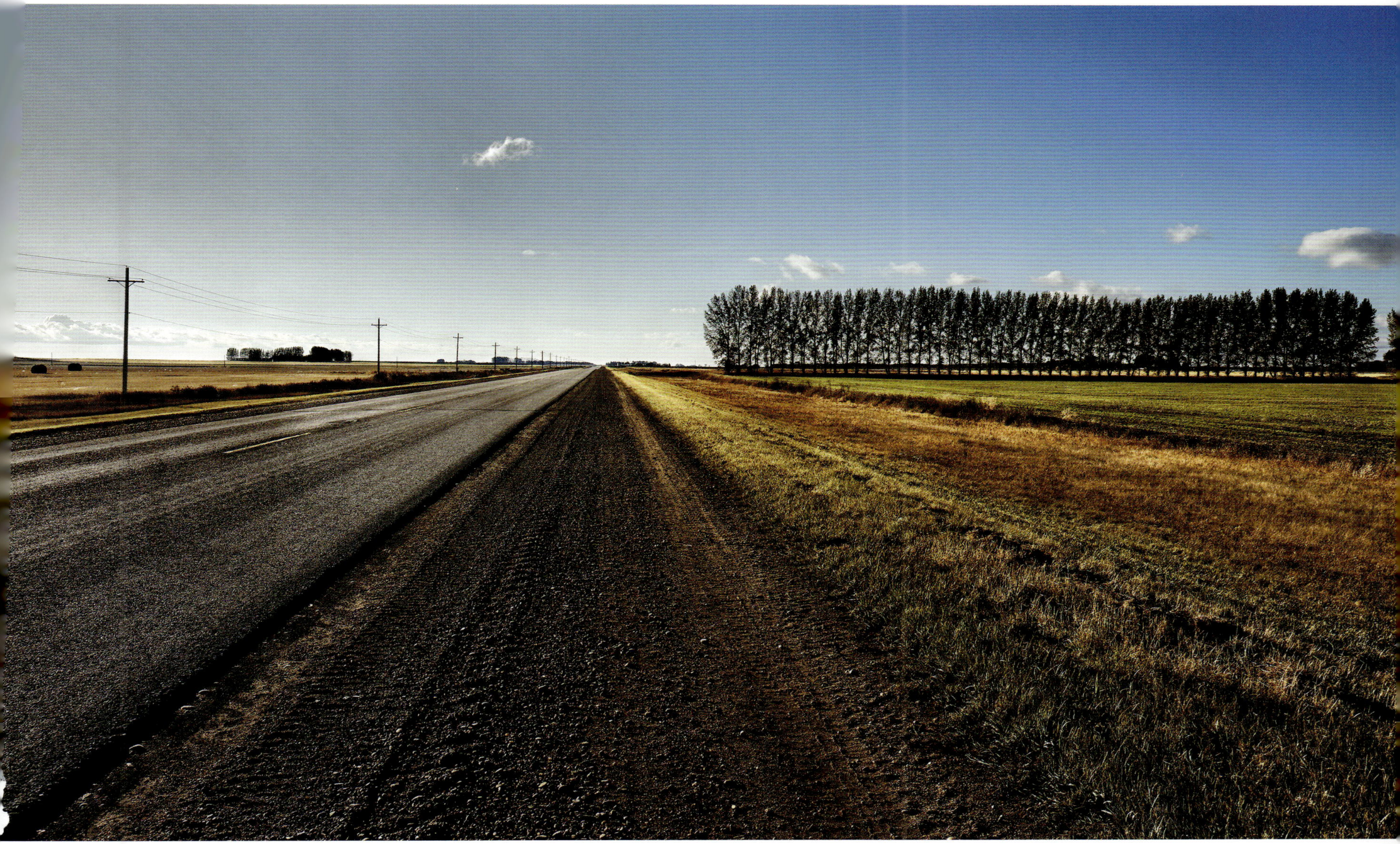

Windbreak, near Souris MB

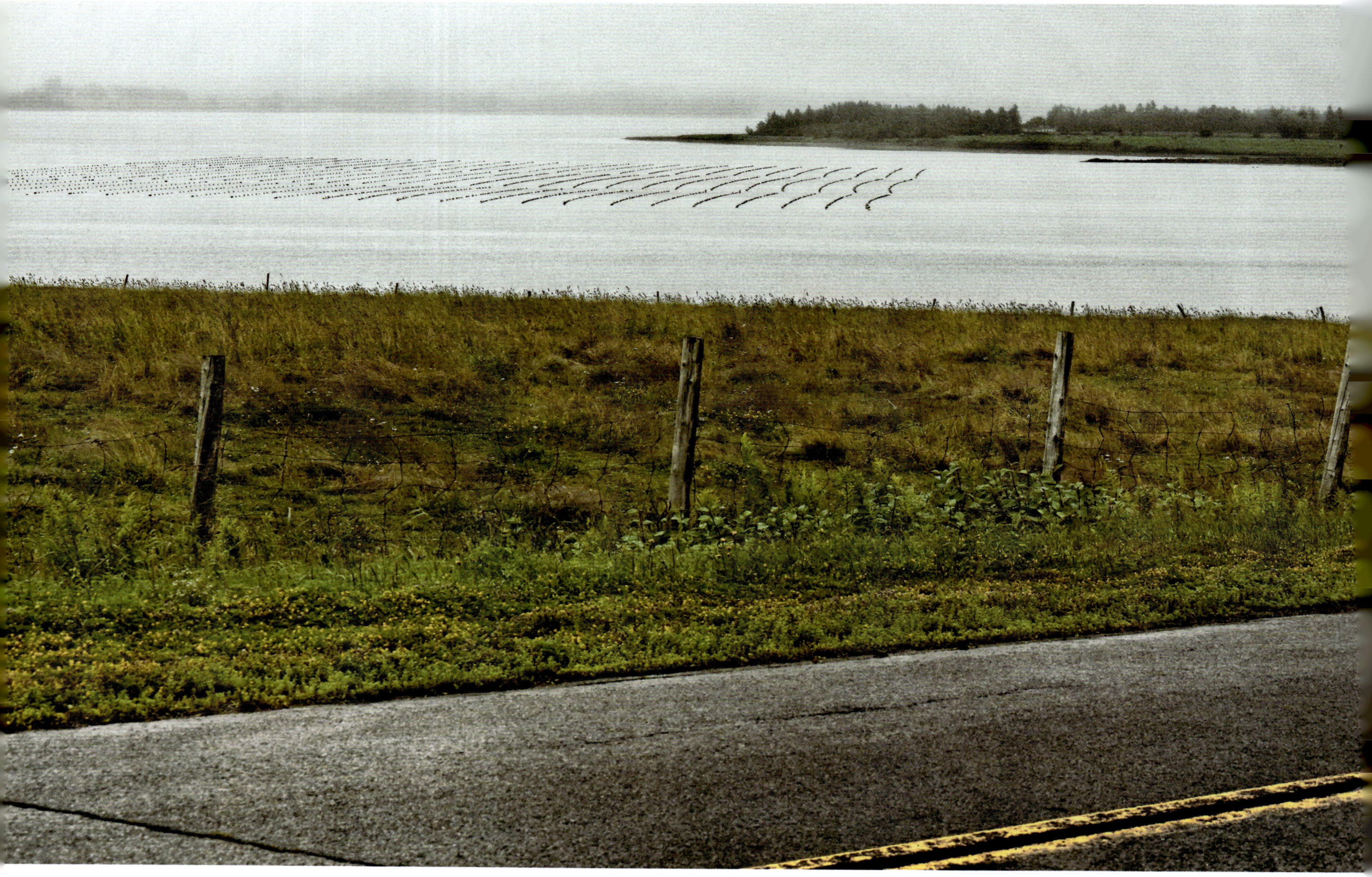

Mussel Farm, Grand River PEI

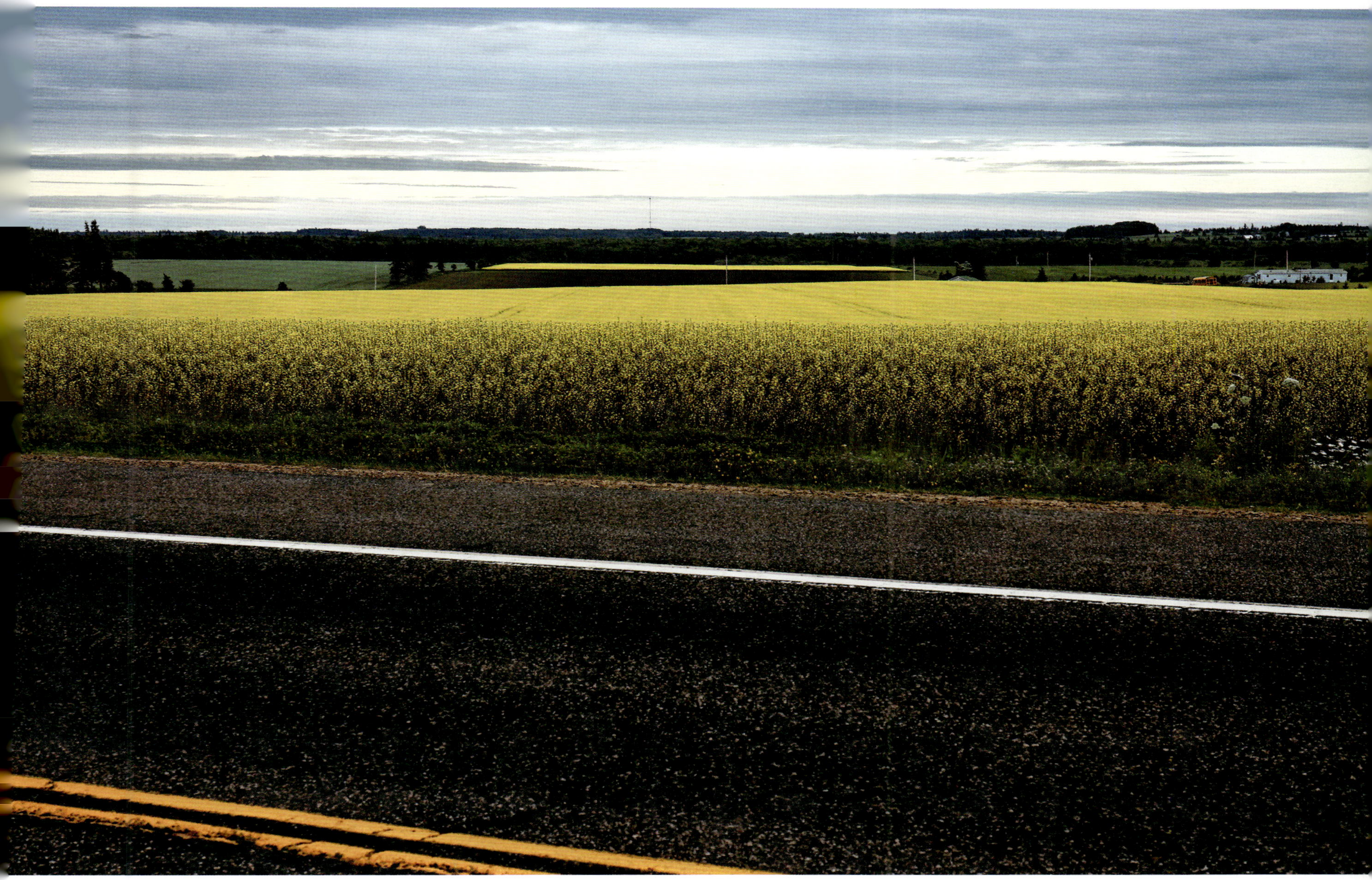

Canola, near Middleton PEI

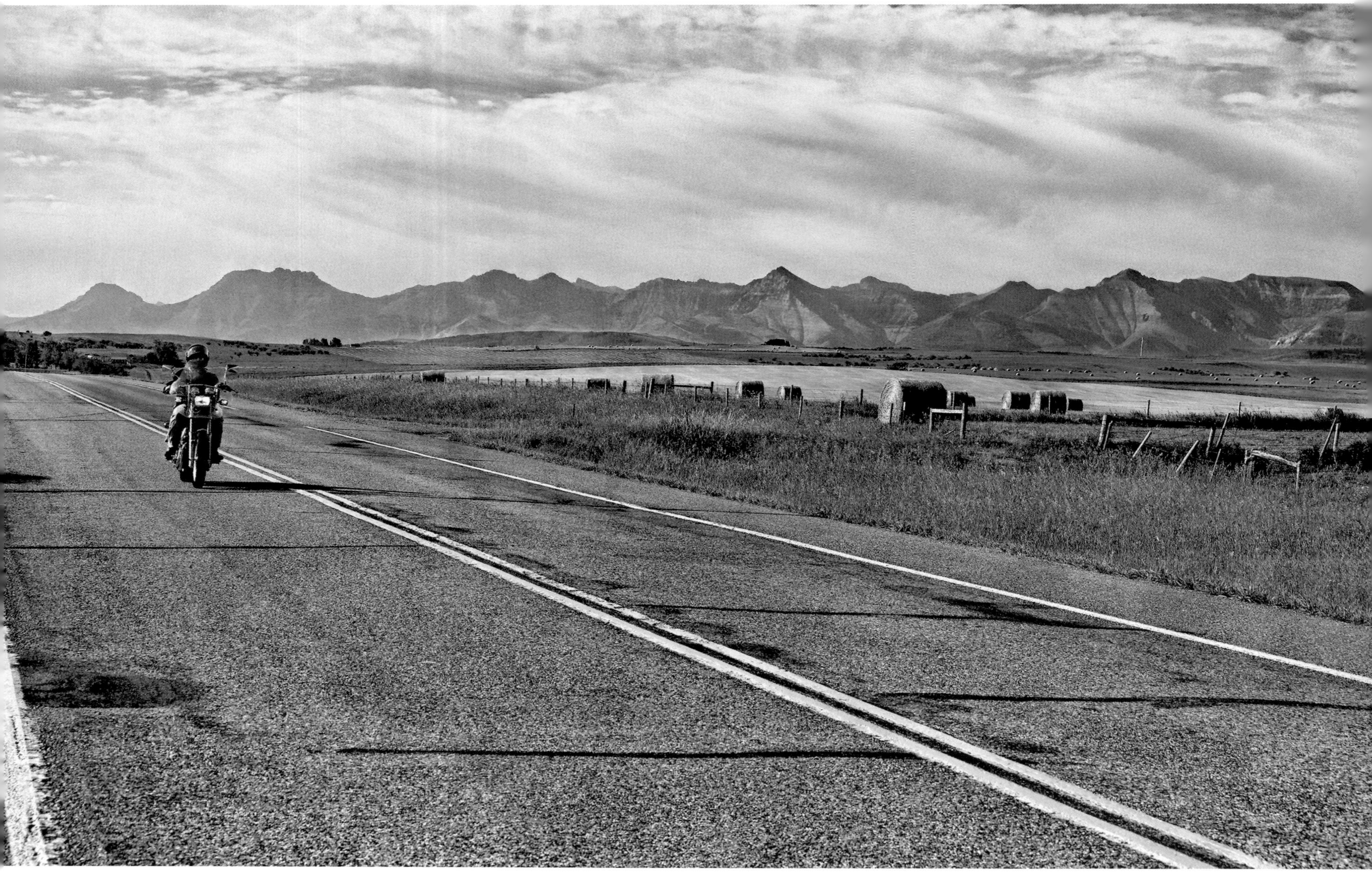

Motorcyclist, Pincher Creek AB

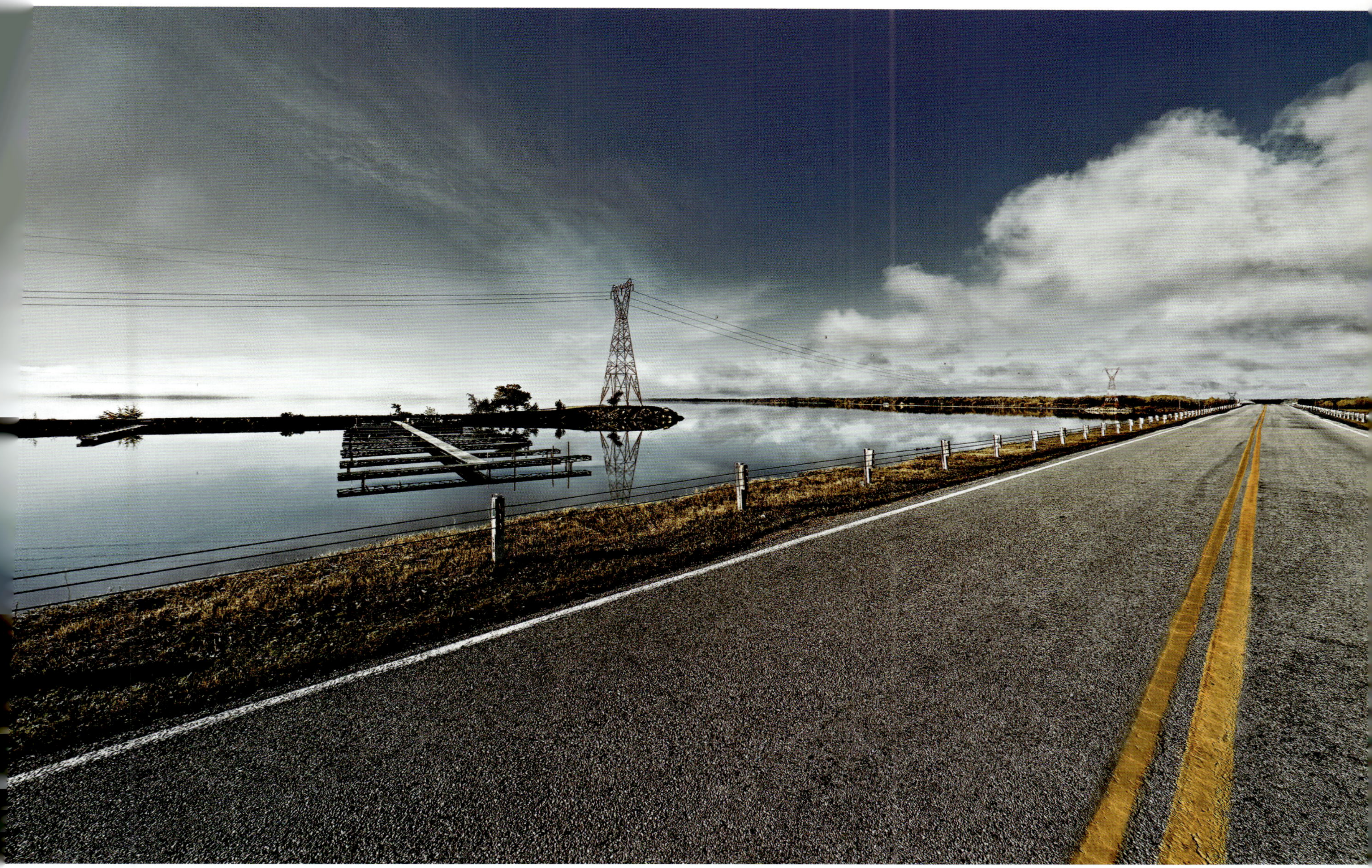

Lake Manitoba Narrows, The Narrows MB

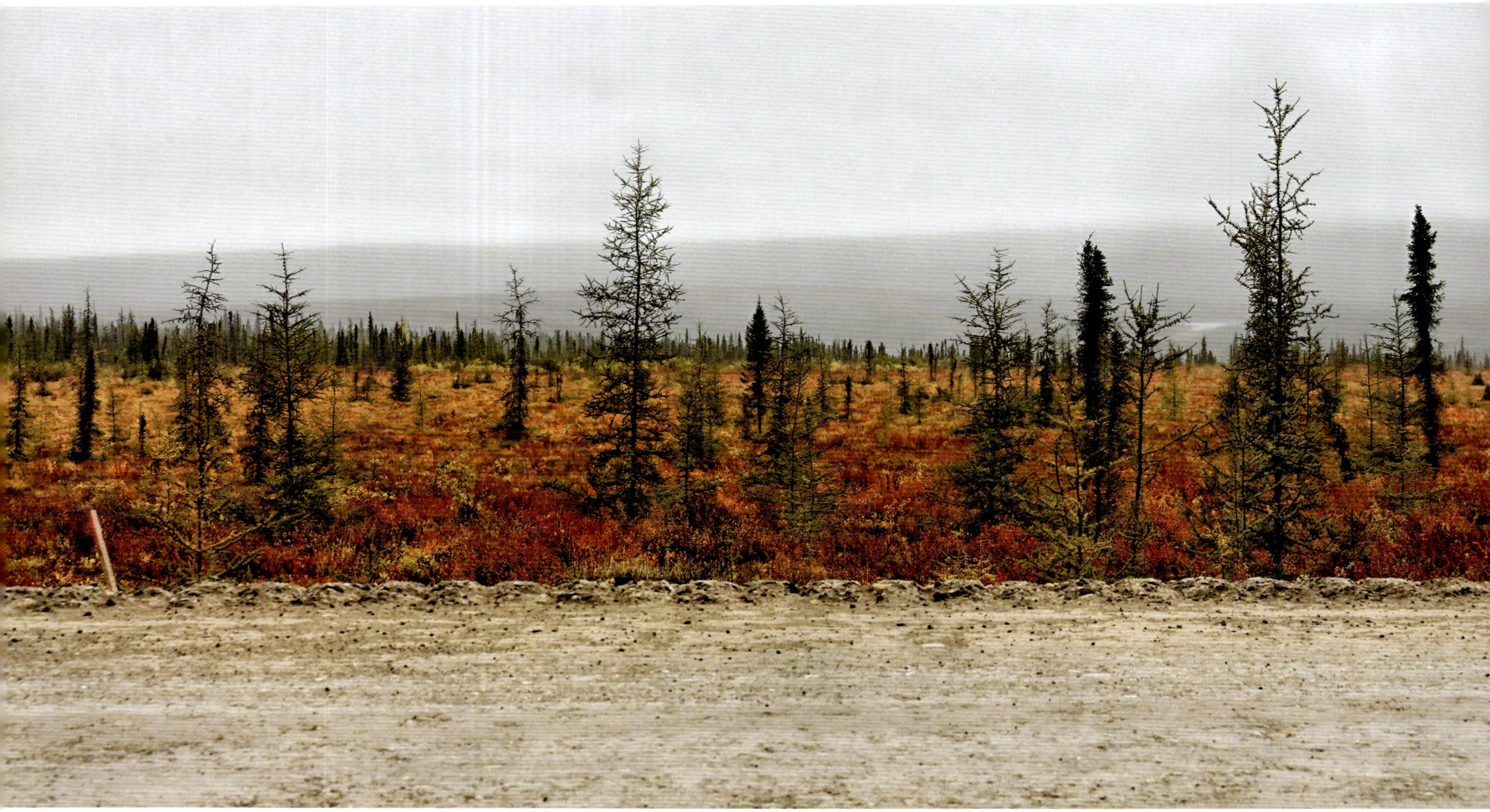

Dempster Highway, near Fort McPherson NT

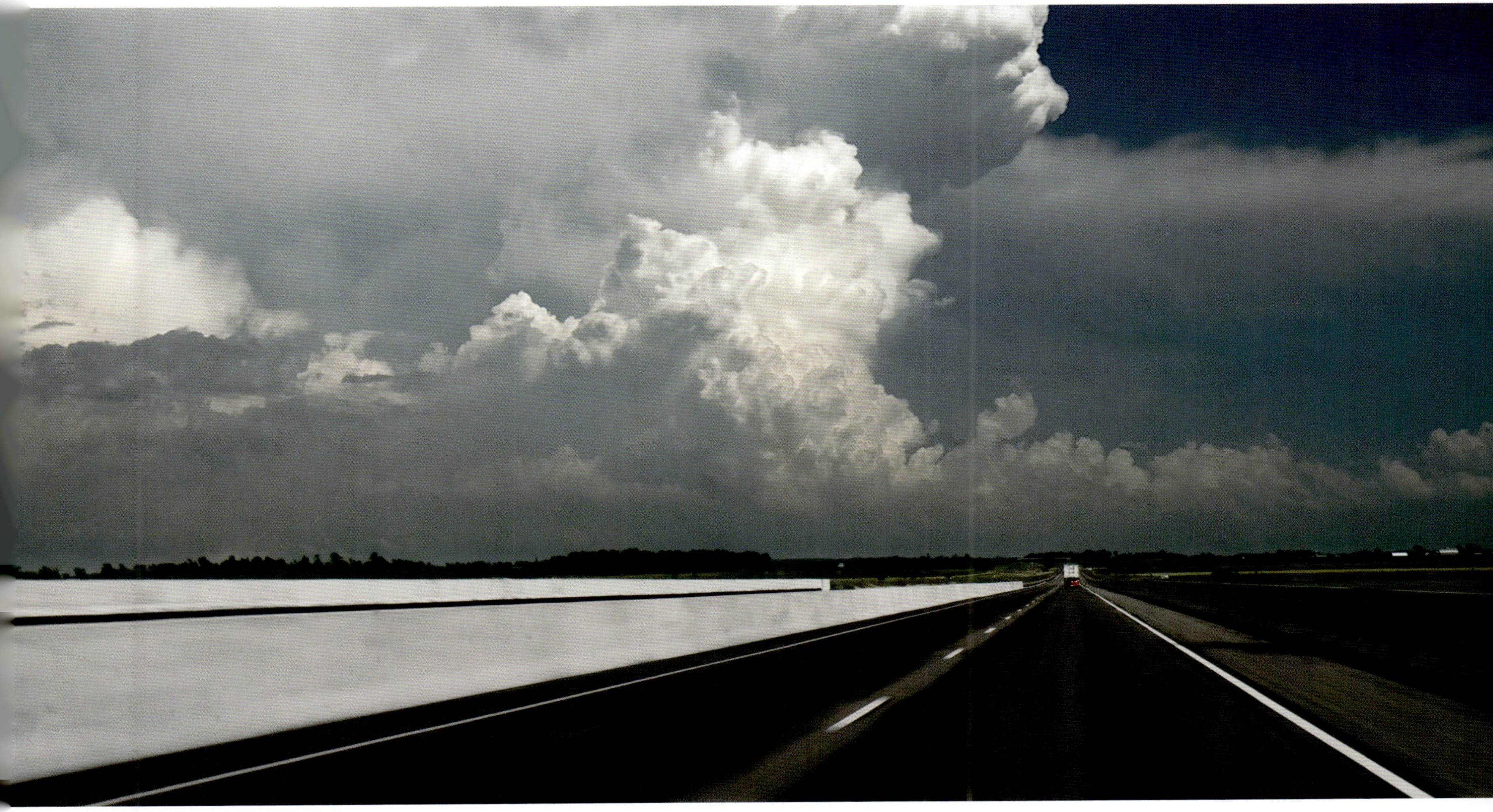

Eastbound, near Arnprior ON

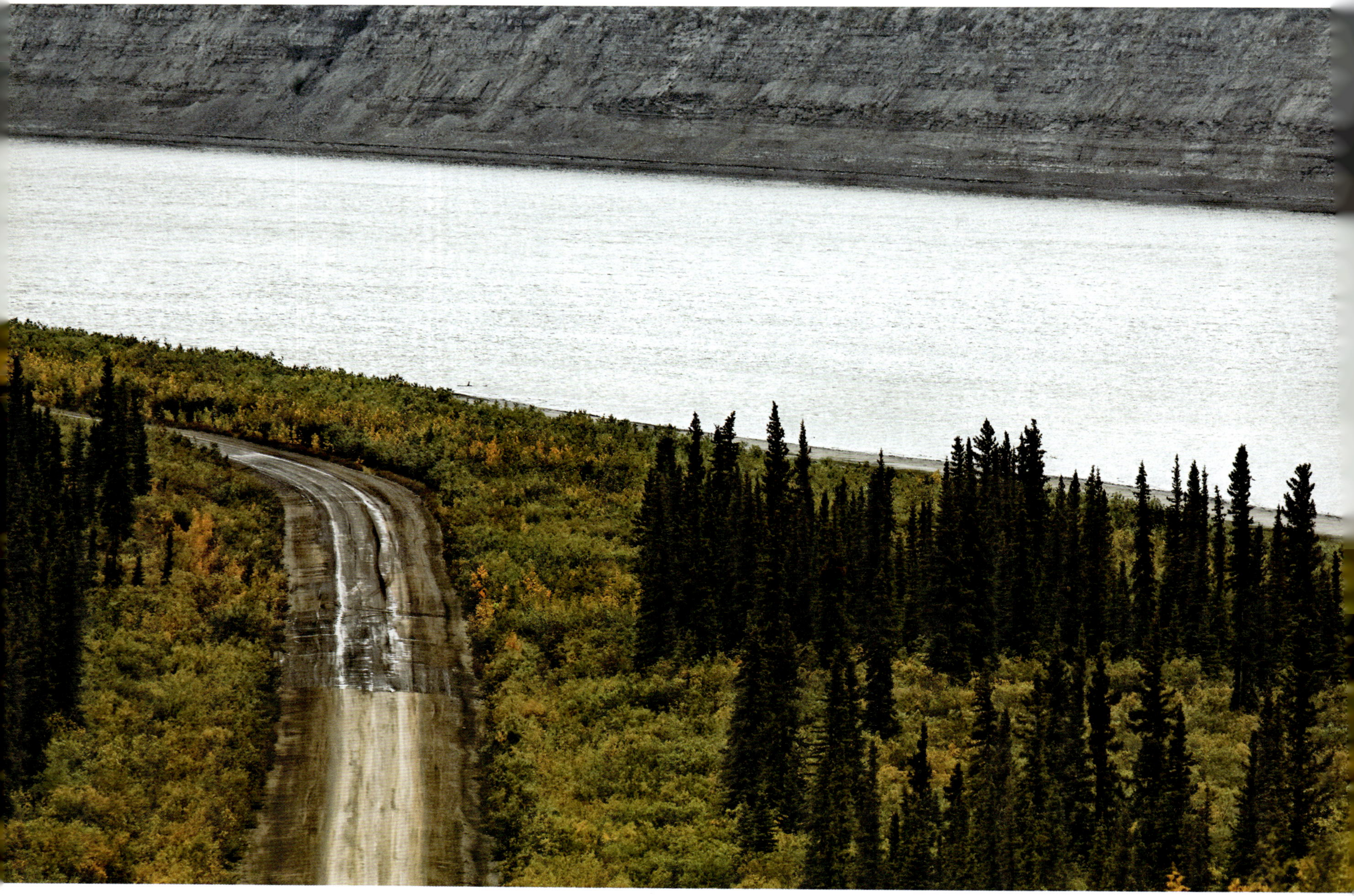

Mackenzie River, near Tsigehtchic NT

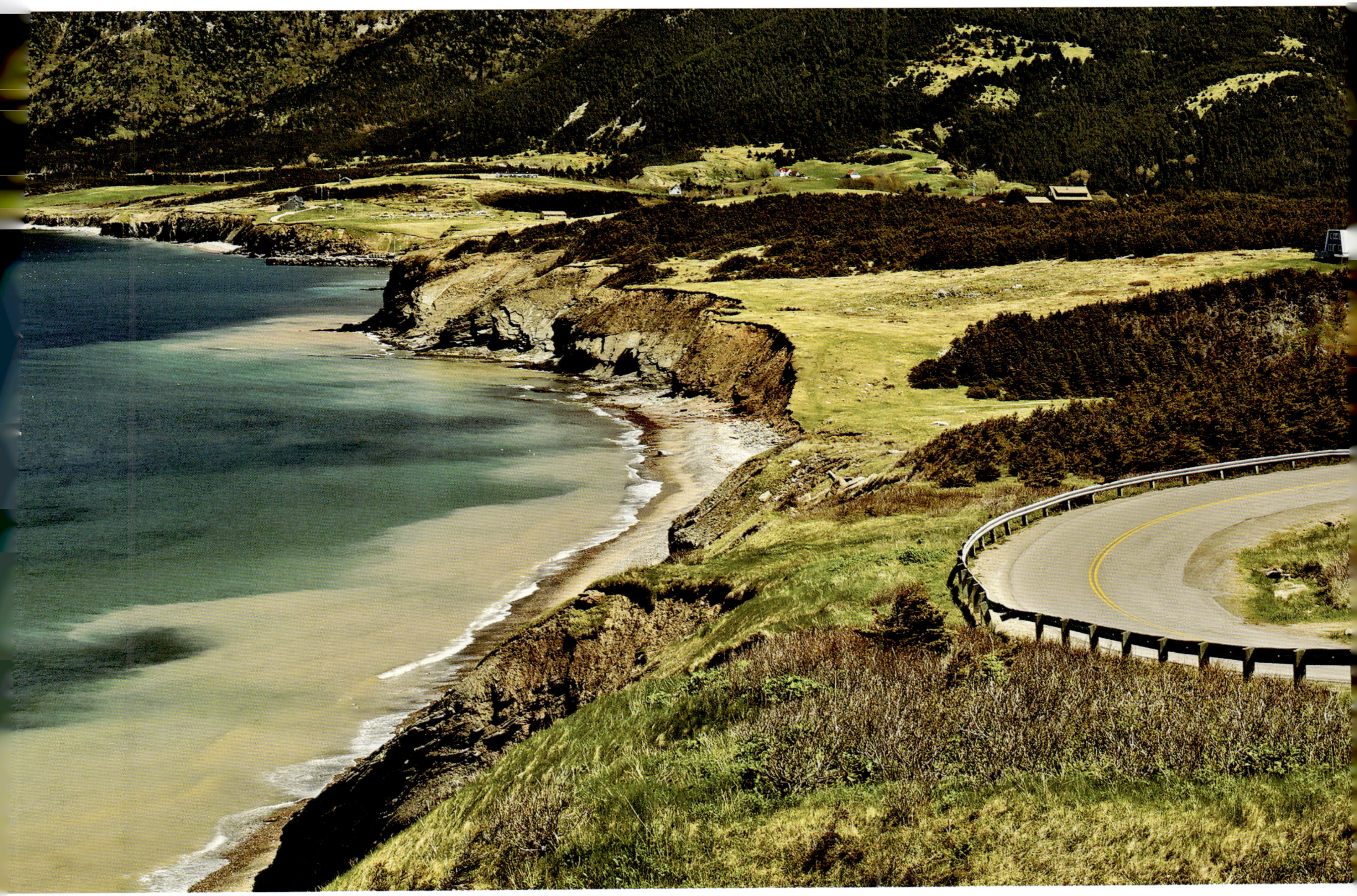

Meat Cove, Meat Cove NS

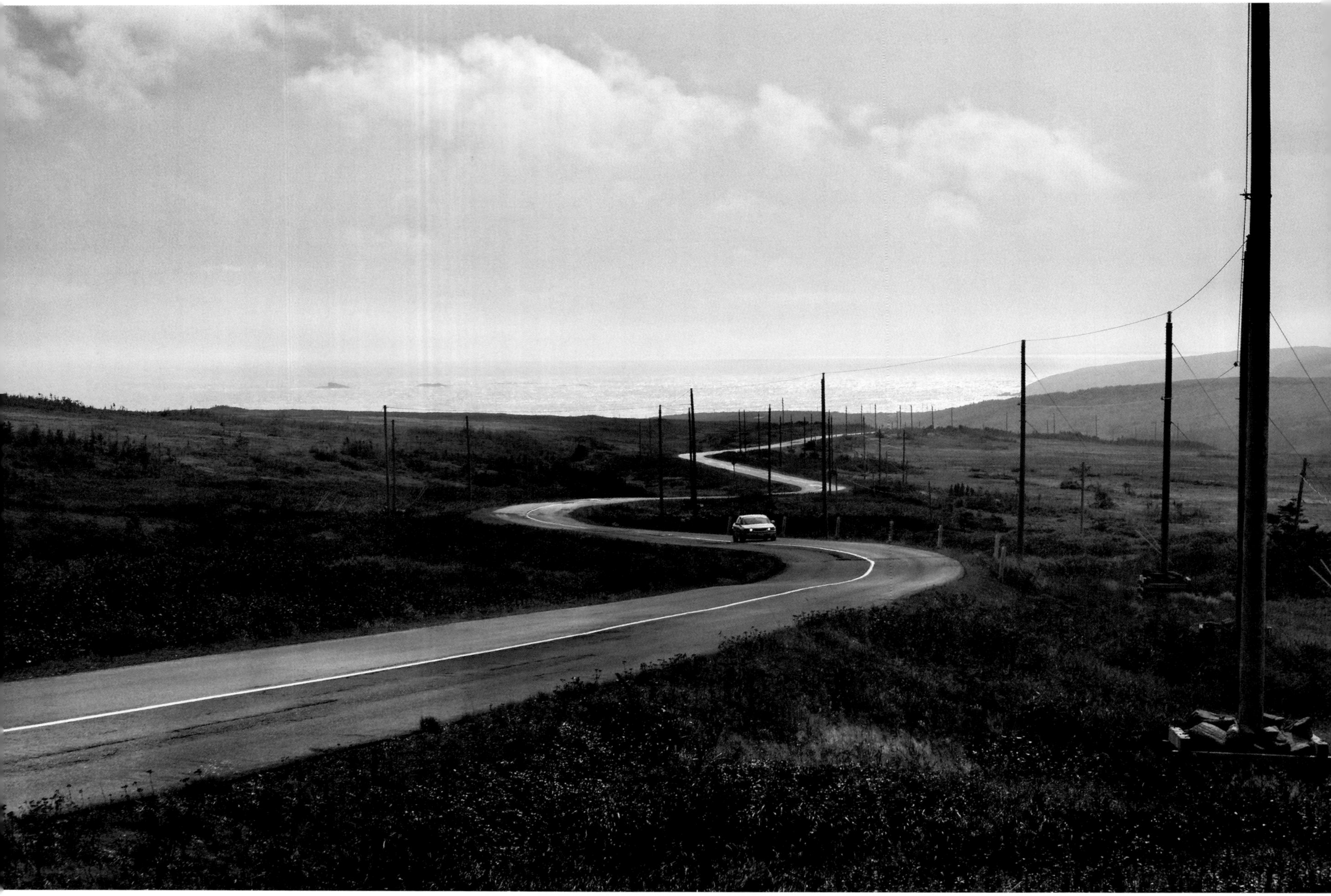

Road to Point Lance, near Point Lance NL

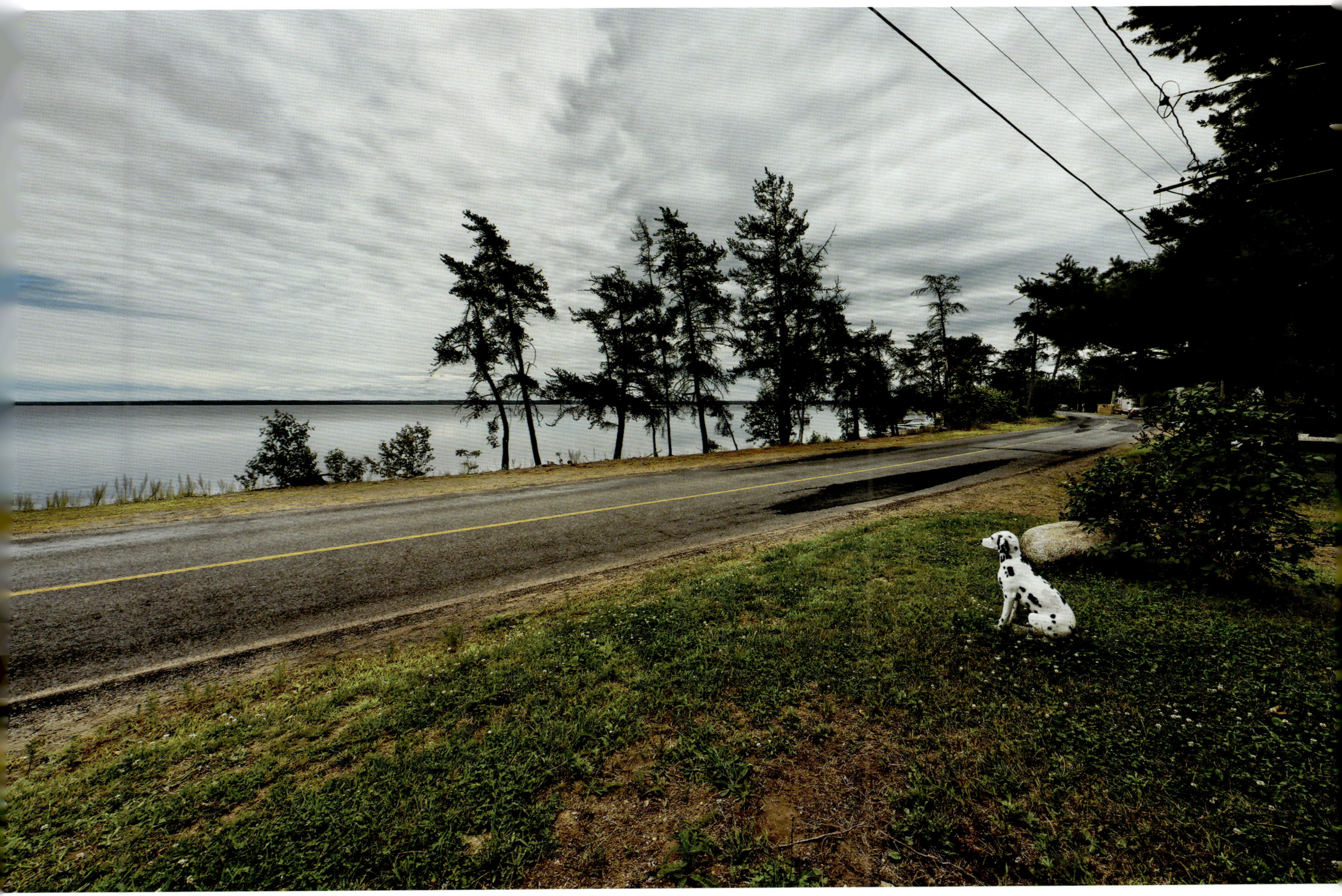

Grand Lake, Princess Park NB

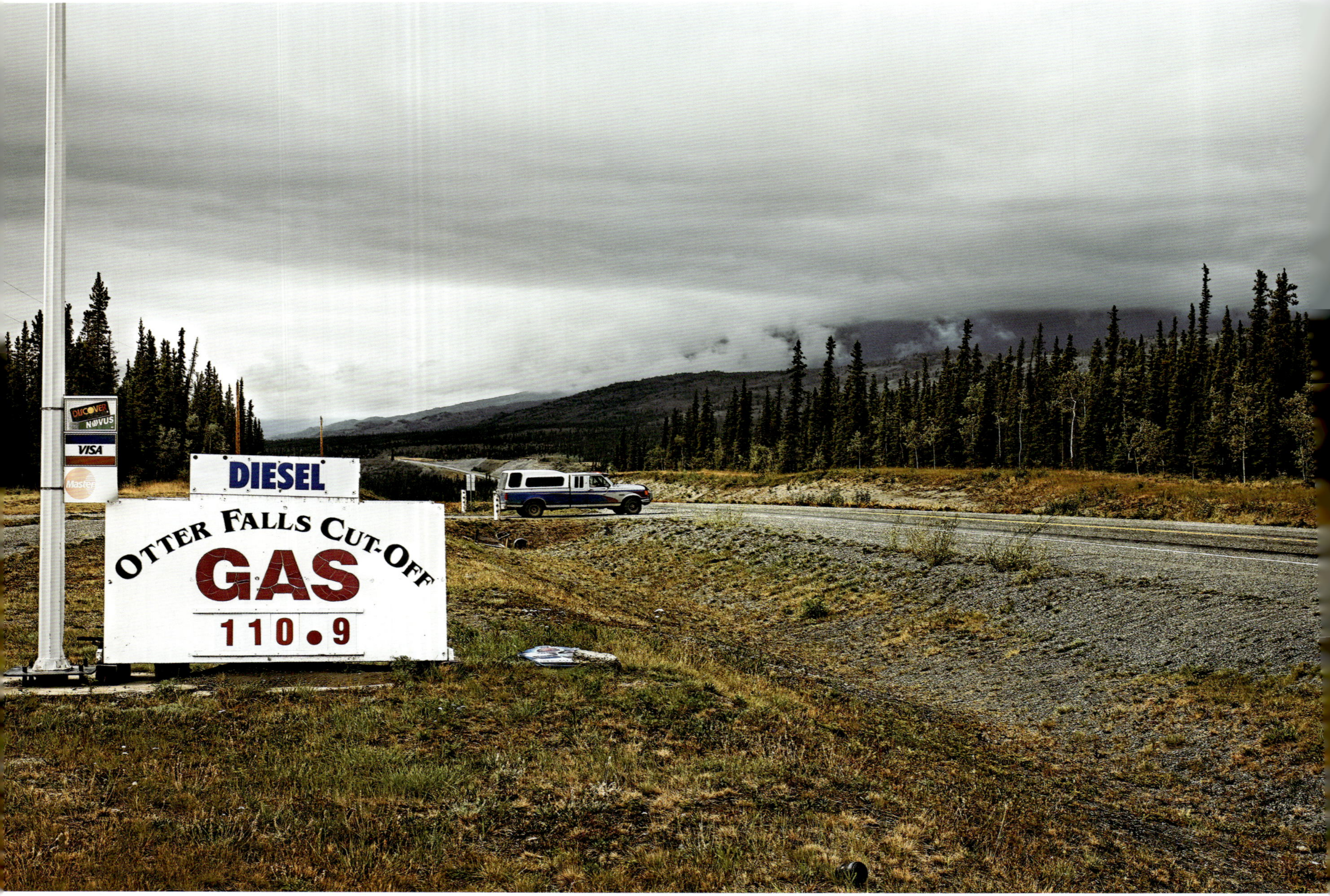

Alaska Highway-Otter Falls Cutoff, east of Haines Junction YK

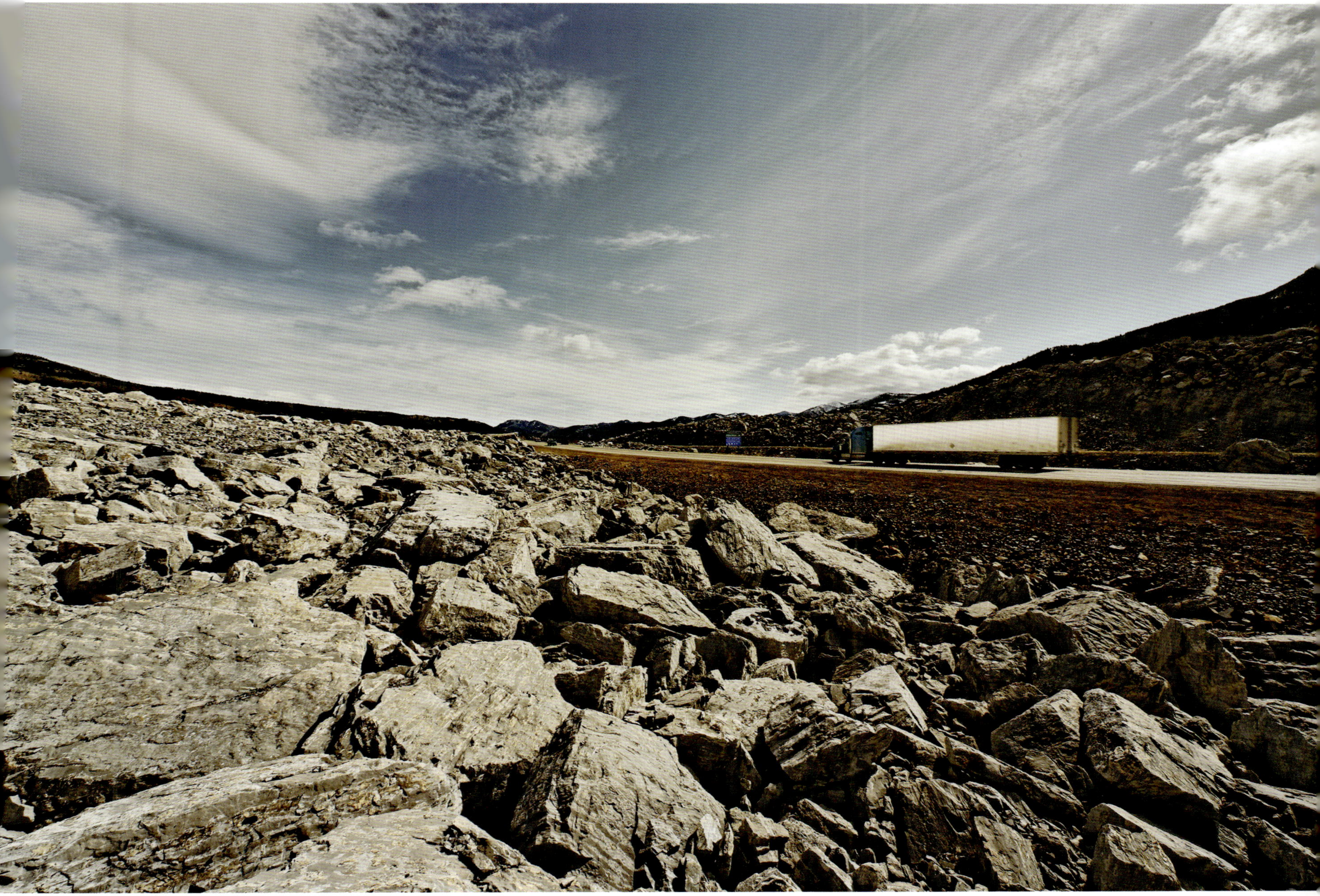

Frank Slide, Frank AB

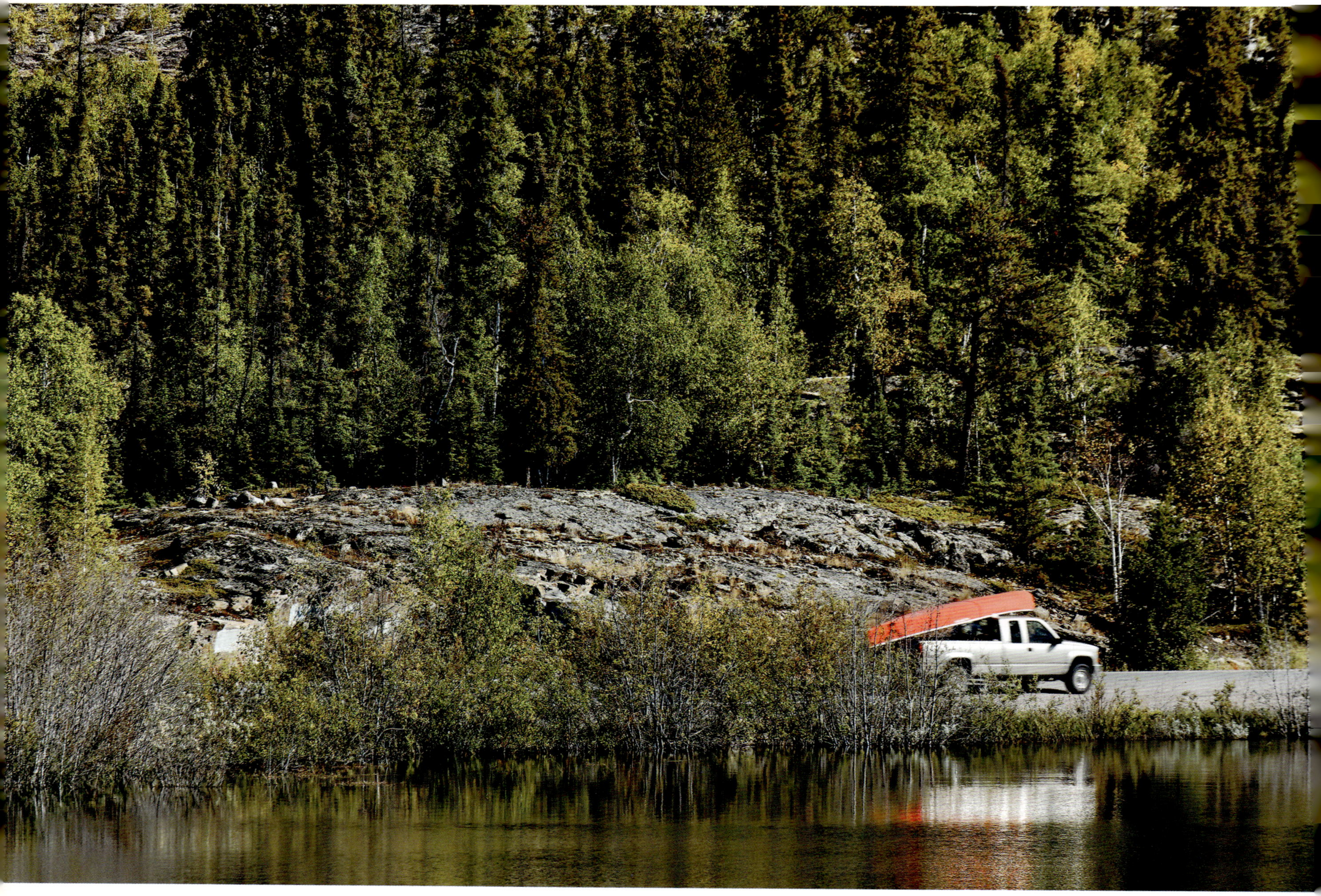

Red Canoe, Ingraham Trail, northeast of Yellowknife NT

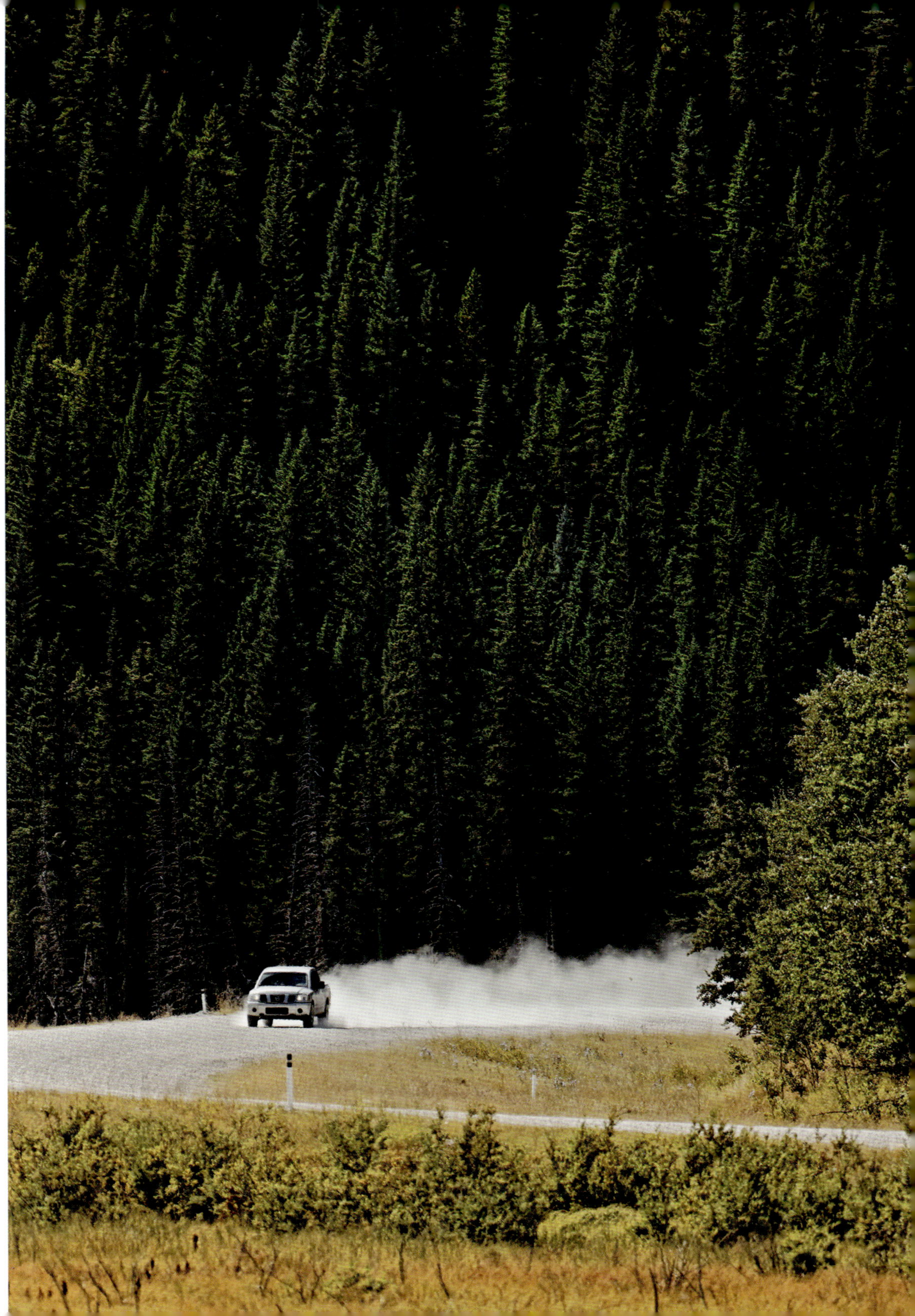

Plume, Kananaskis Valley AB

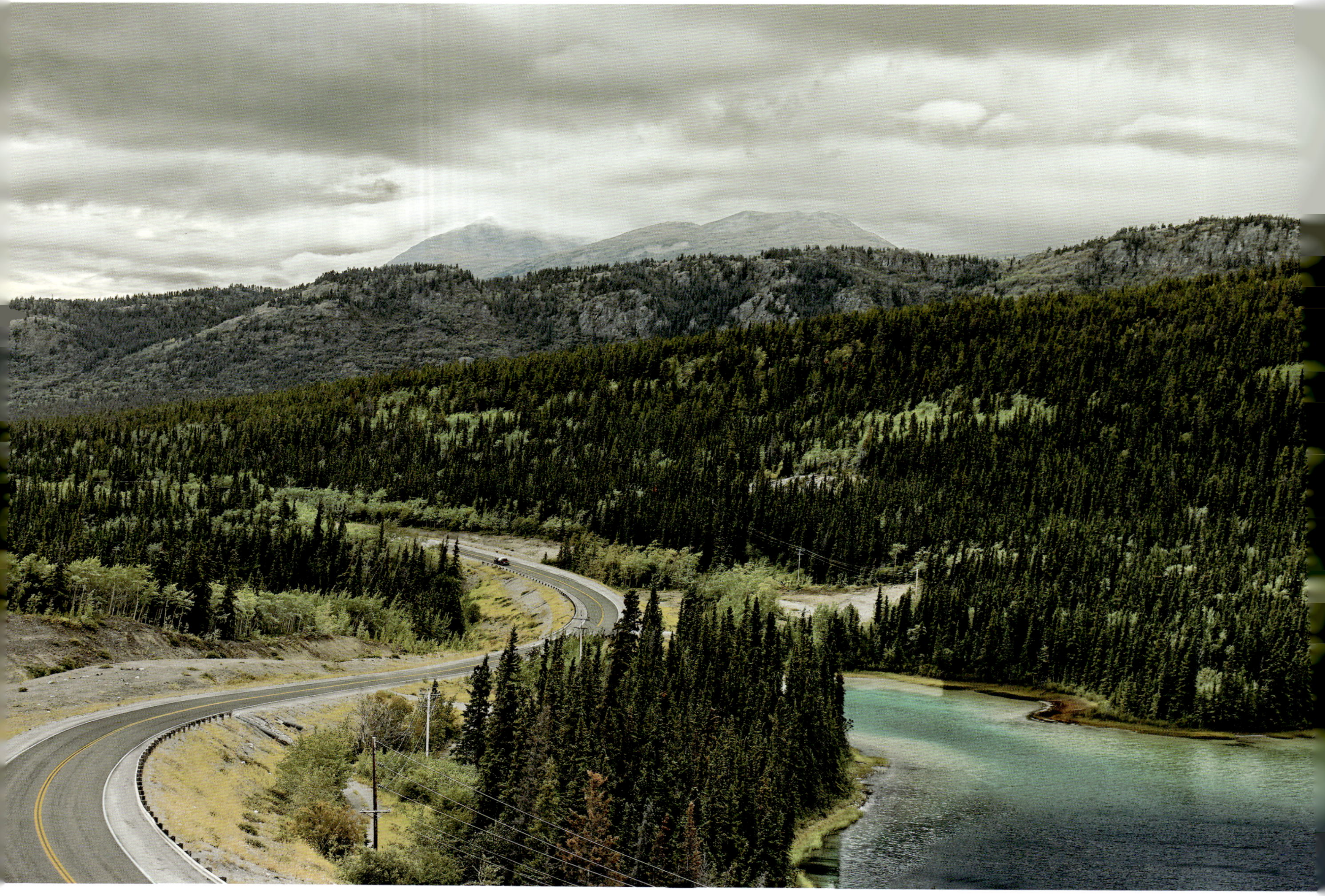

Emerald Lake, near Carcross YK

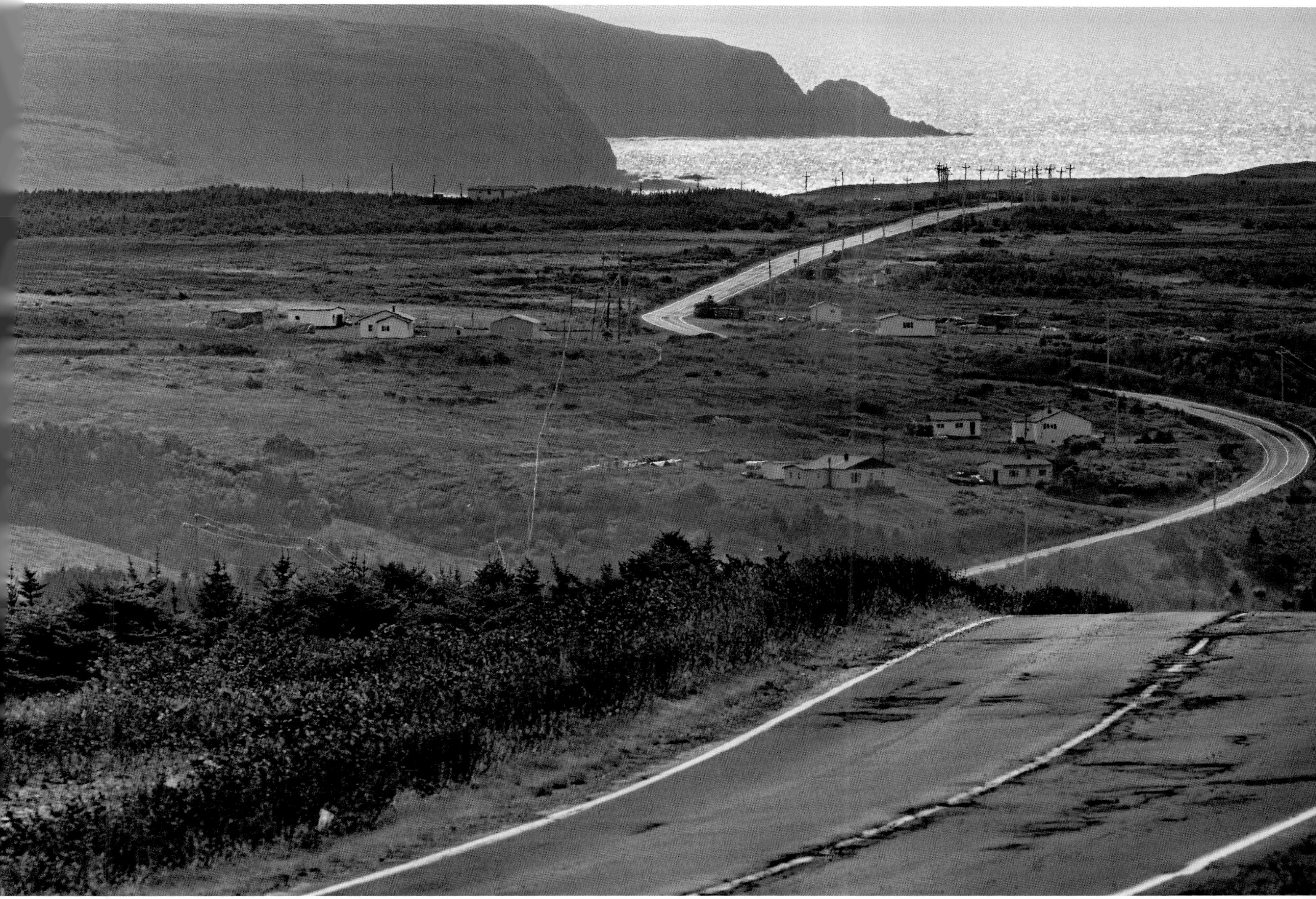

Looking South from Angel's Cove, Angel's Cove NL

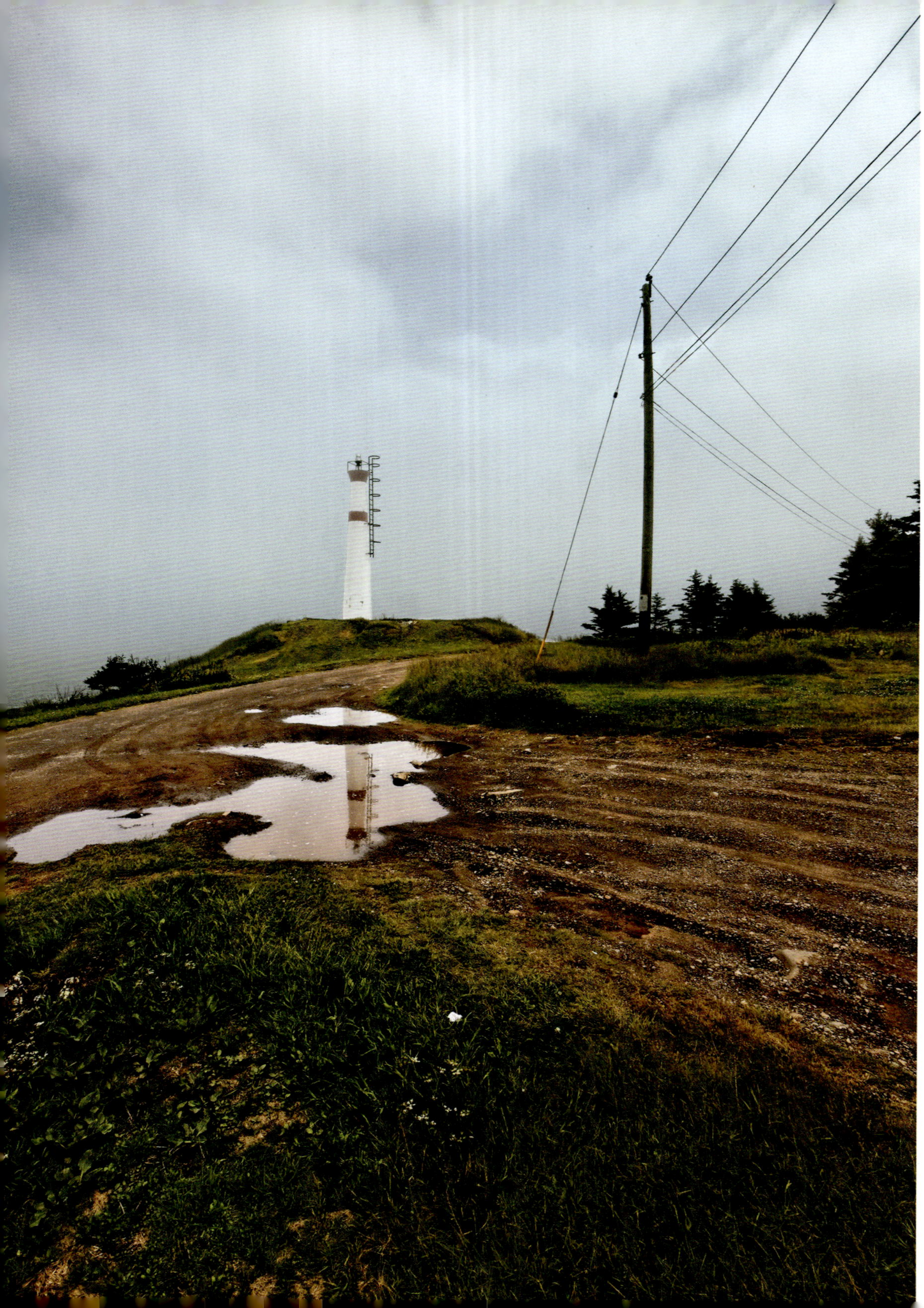

Puddle, Black Rock NS

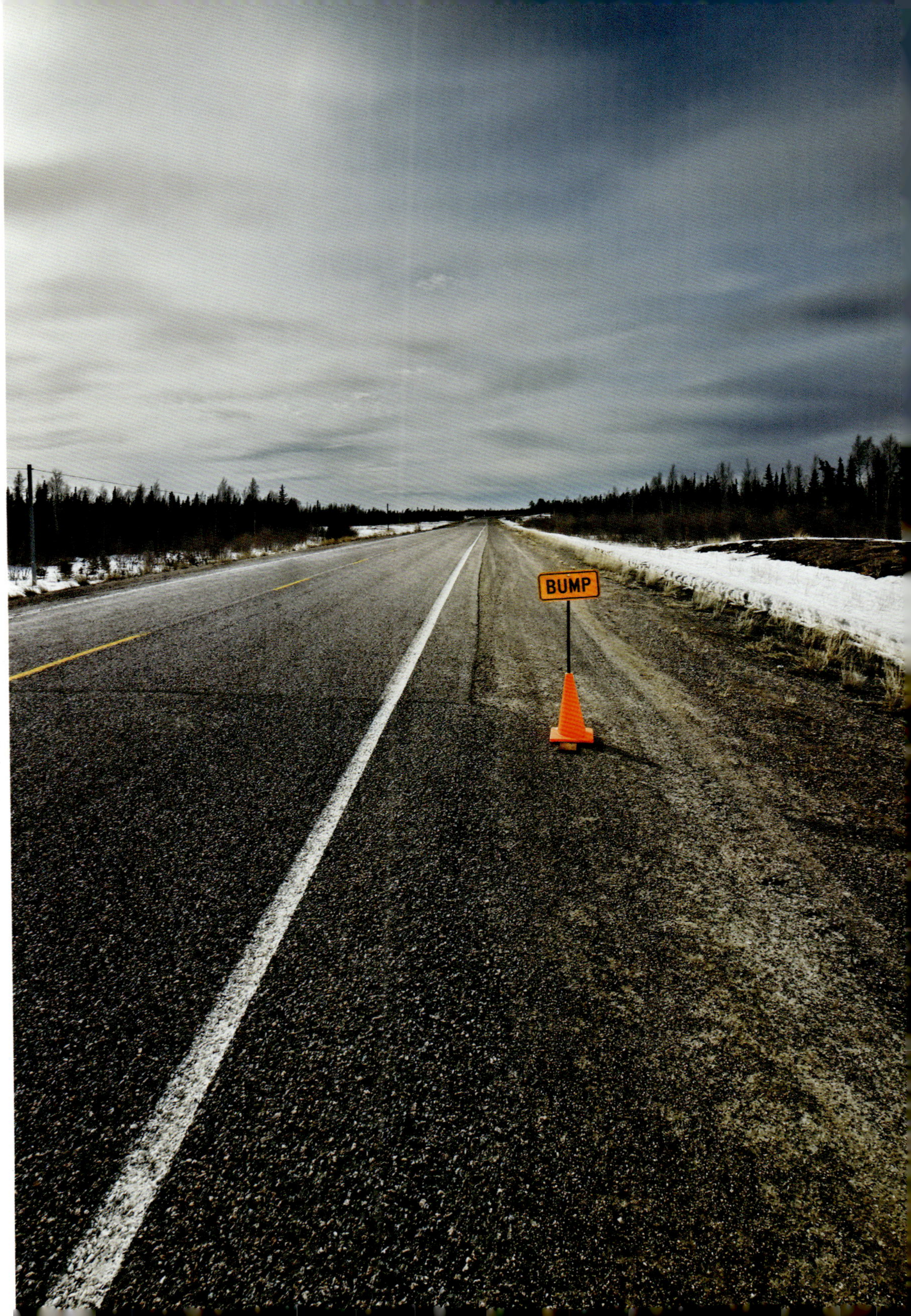

Bump, Highway 3, west of Yellowknife NT

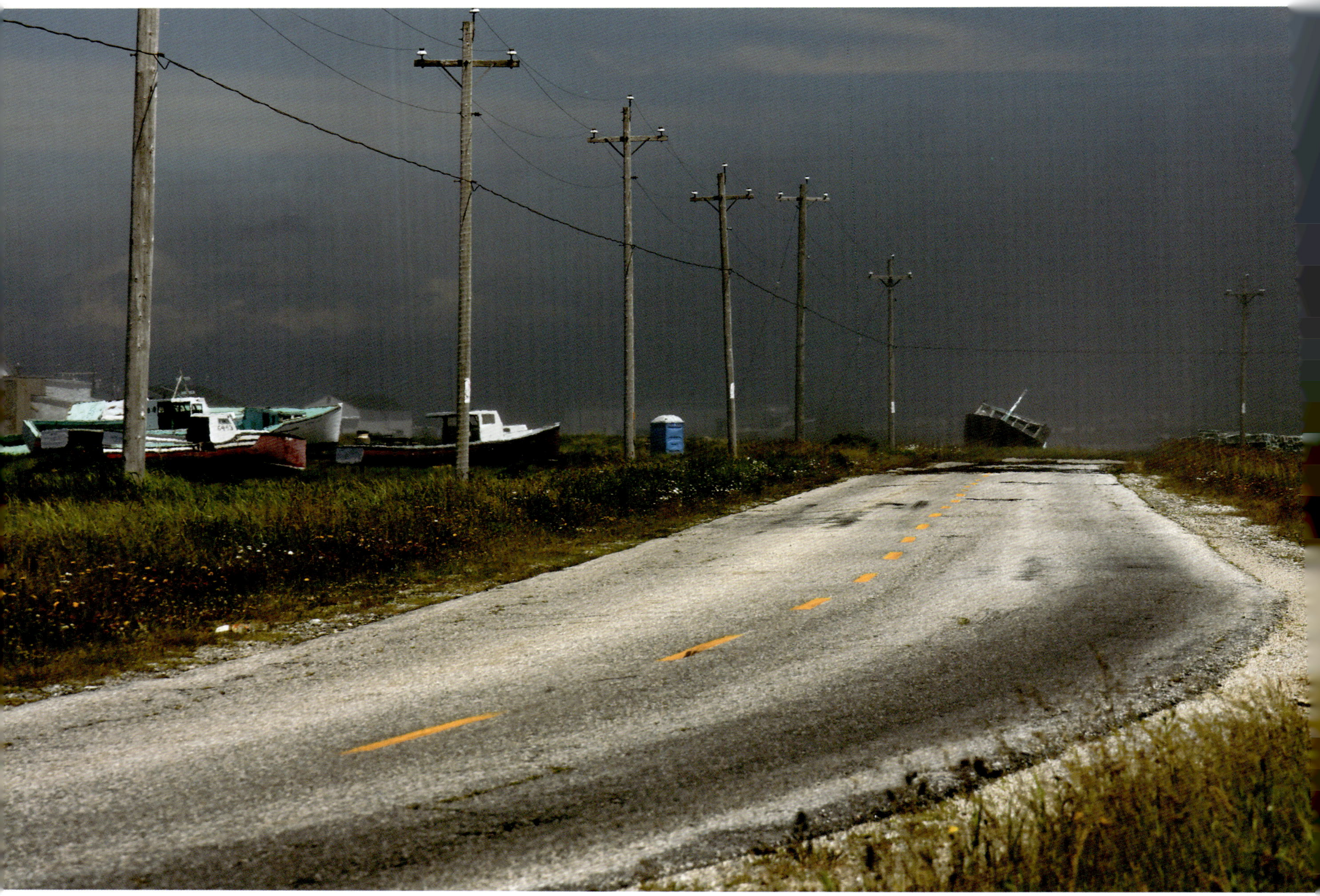

Beached, Daniels Head NS

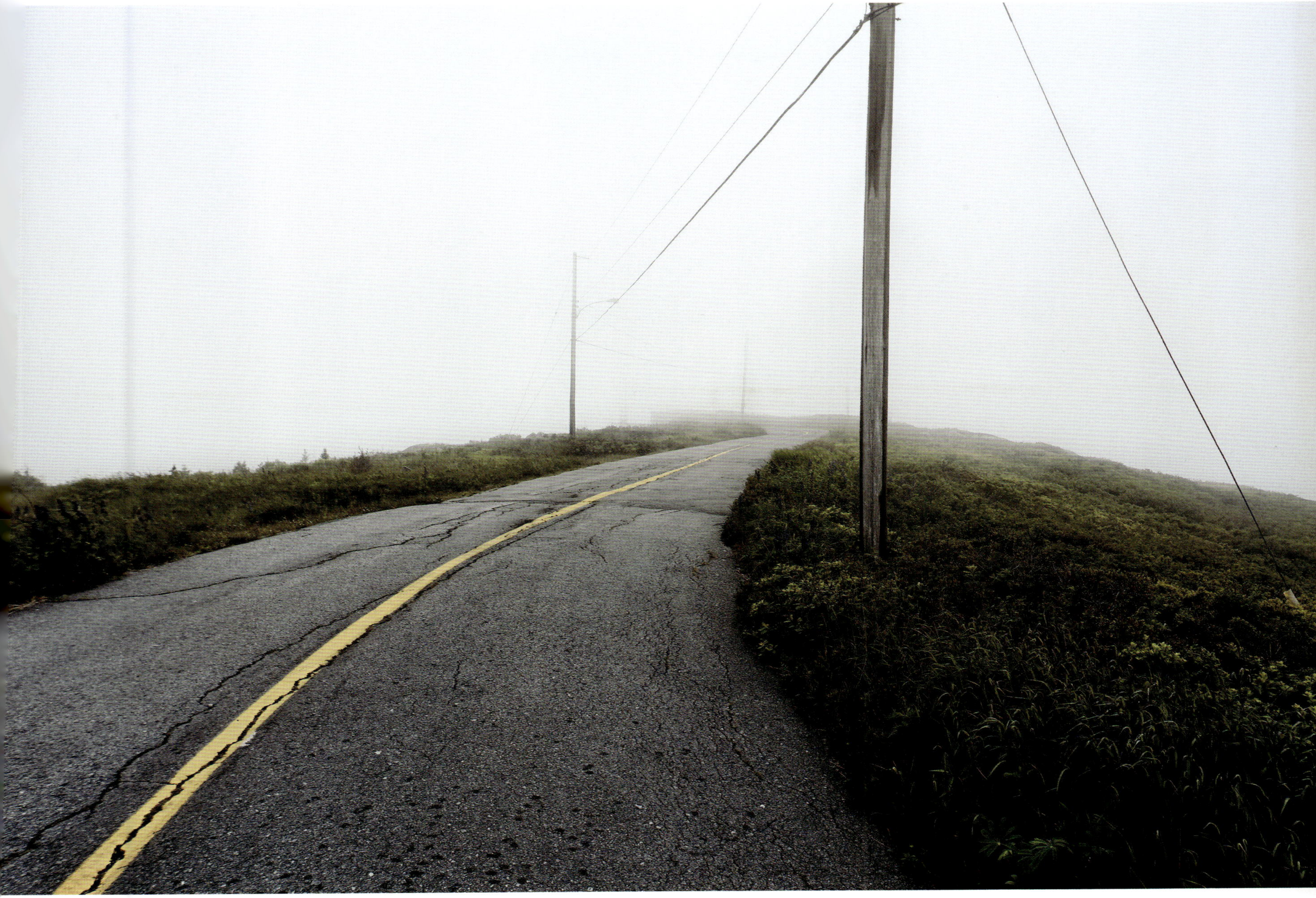

Chebucto Head, near Ketch Harbour NS

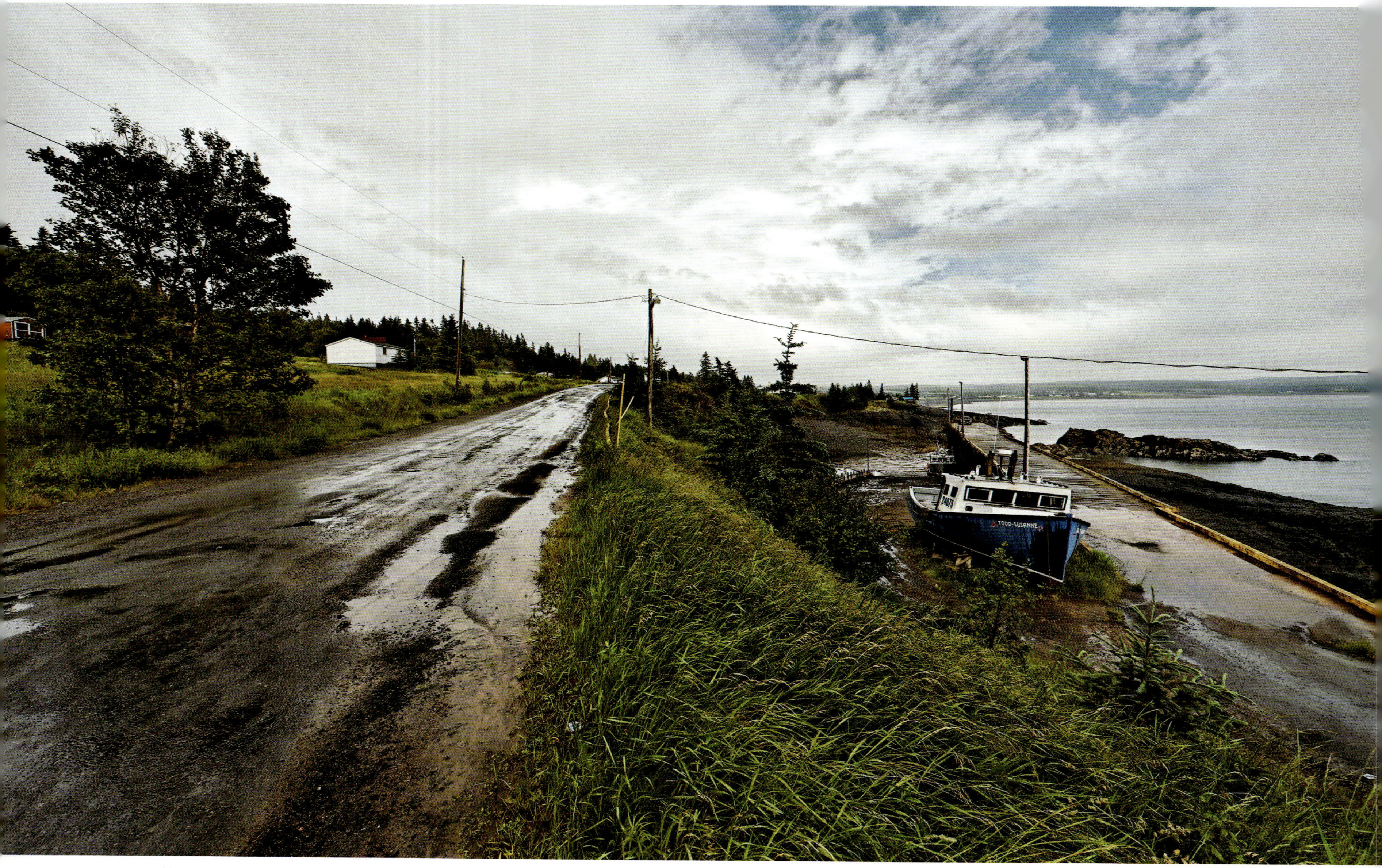

Boat on the Road, Scots Bay NS

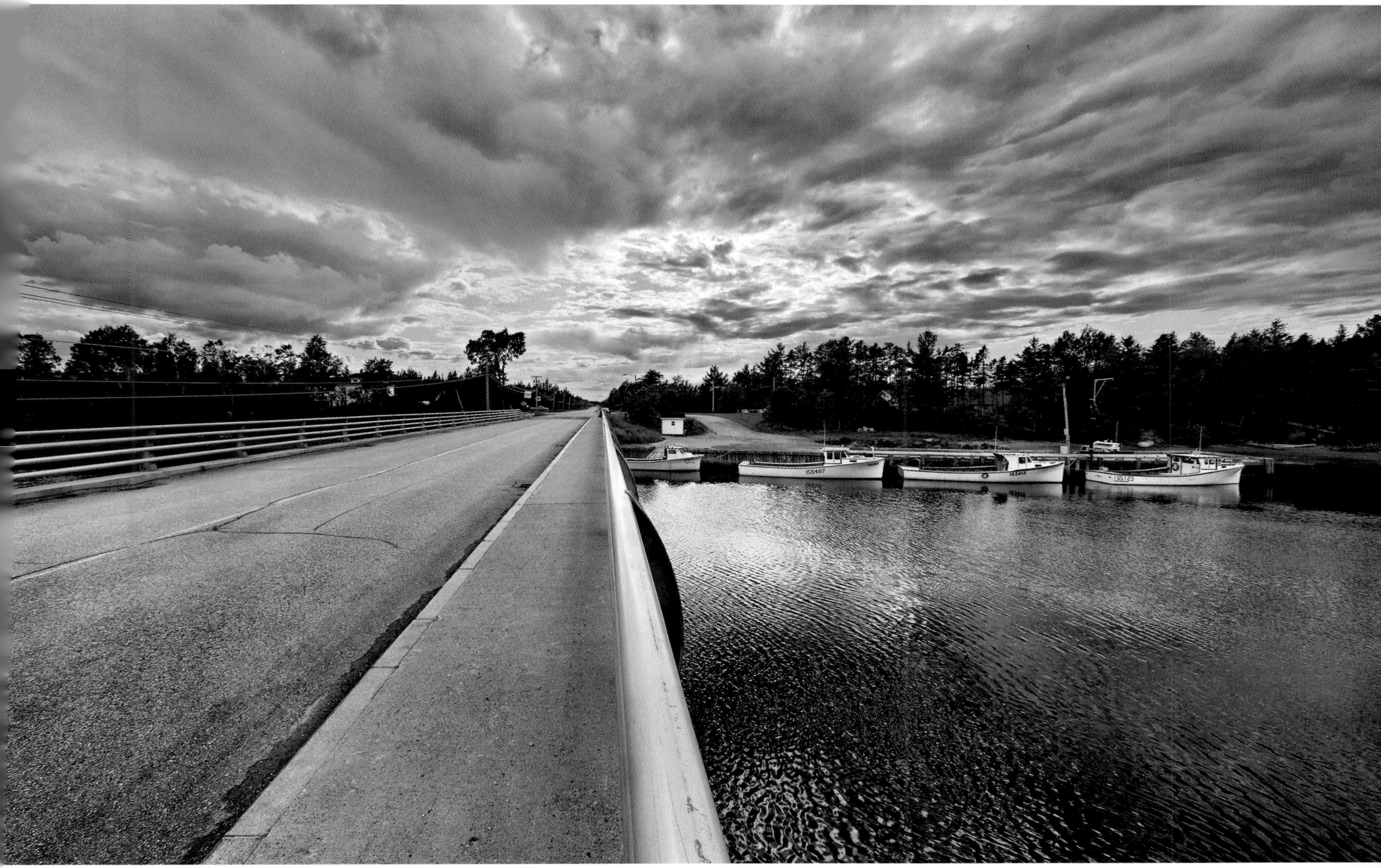

Three-and-a-Half-Boats, Bay du Vin NB

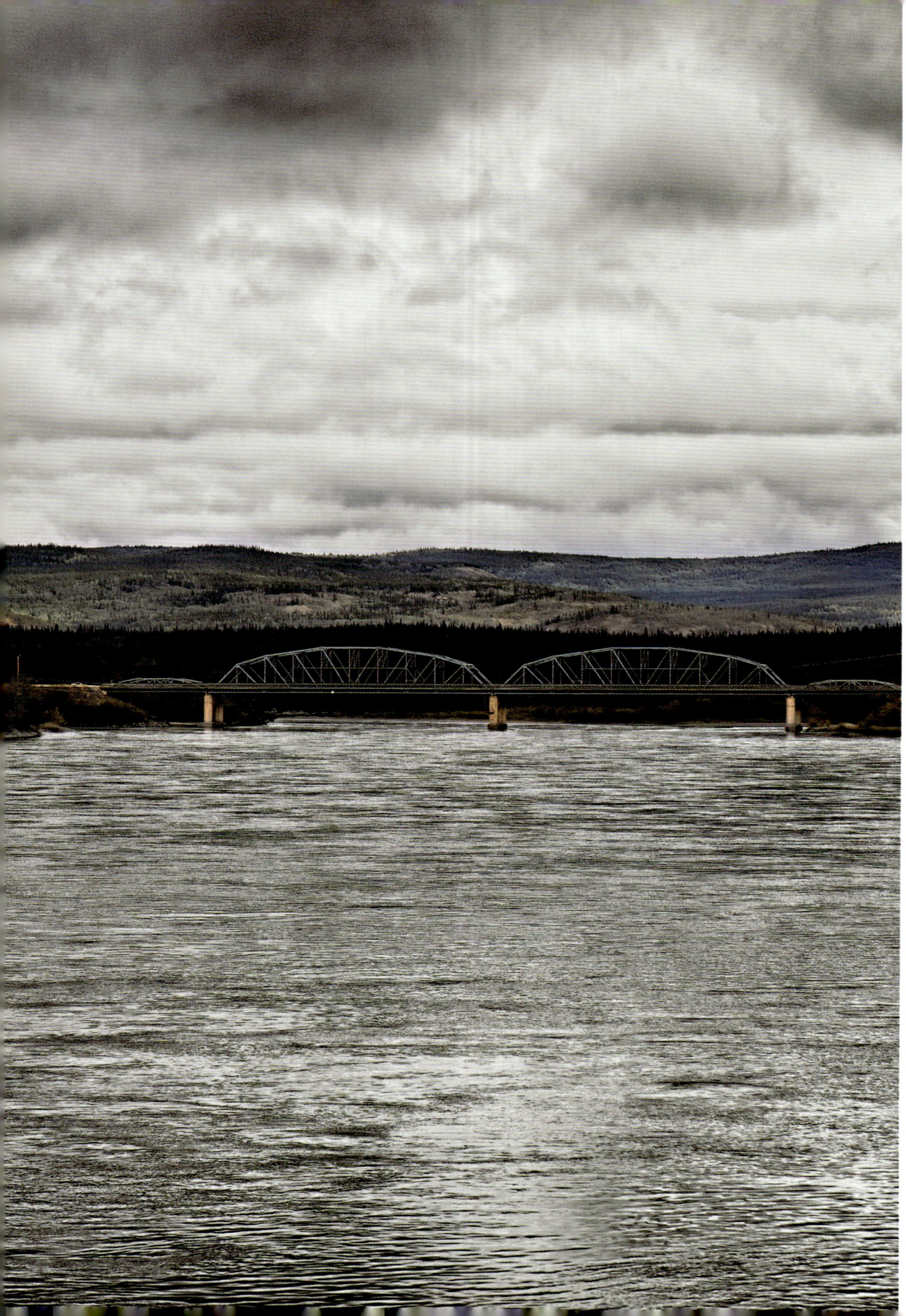

Yukon River Bridge, Carmacks YK

Confederation Bridge, North Carleton PEI

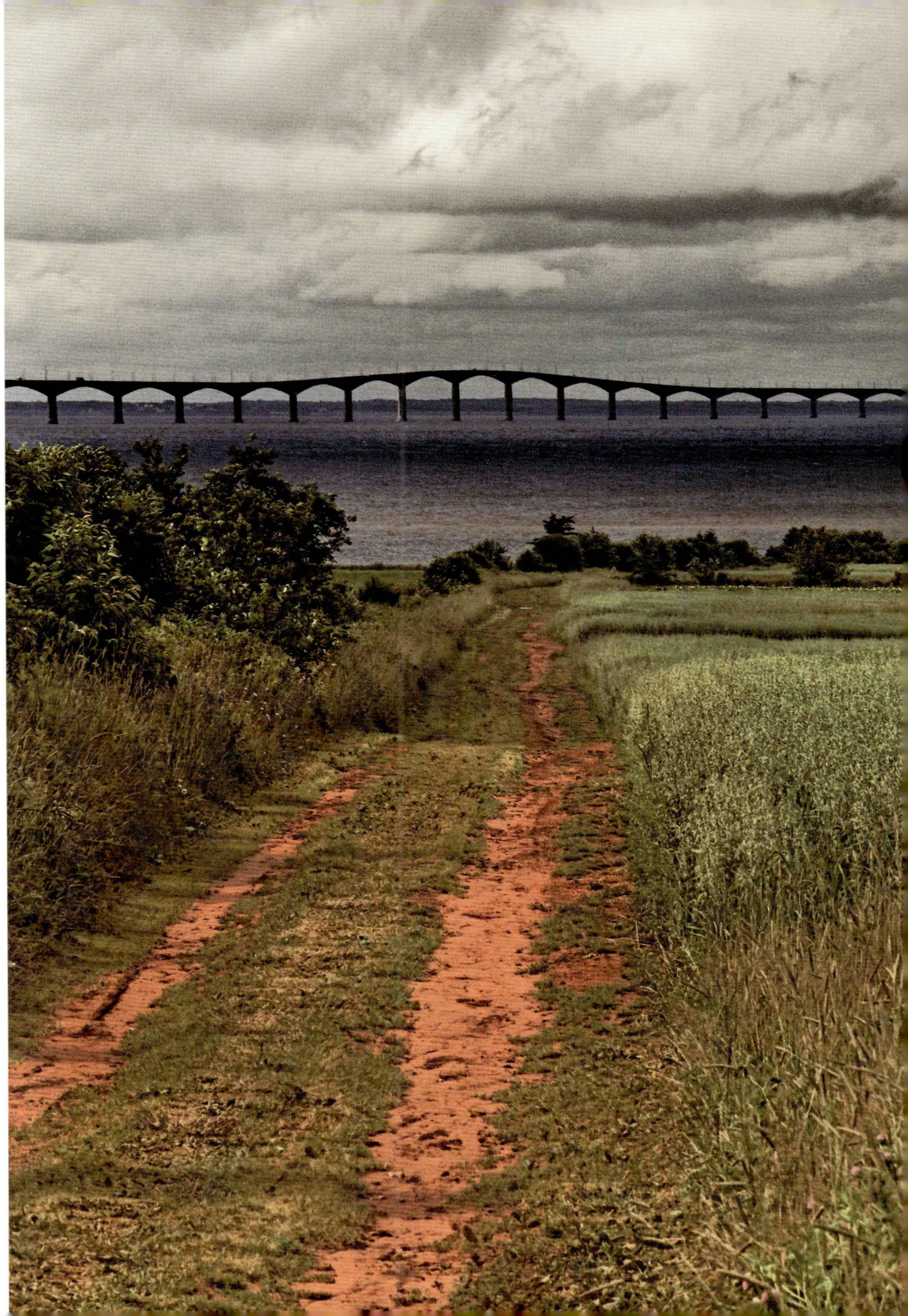

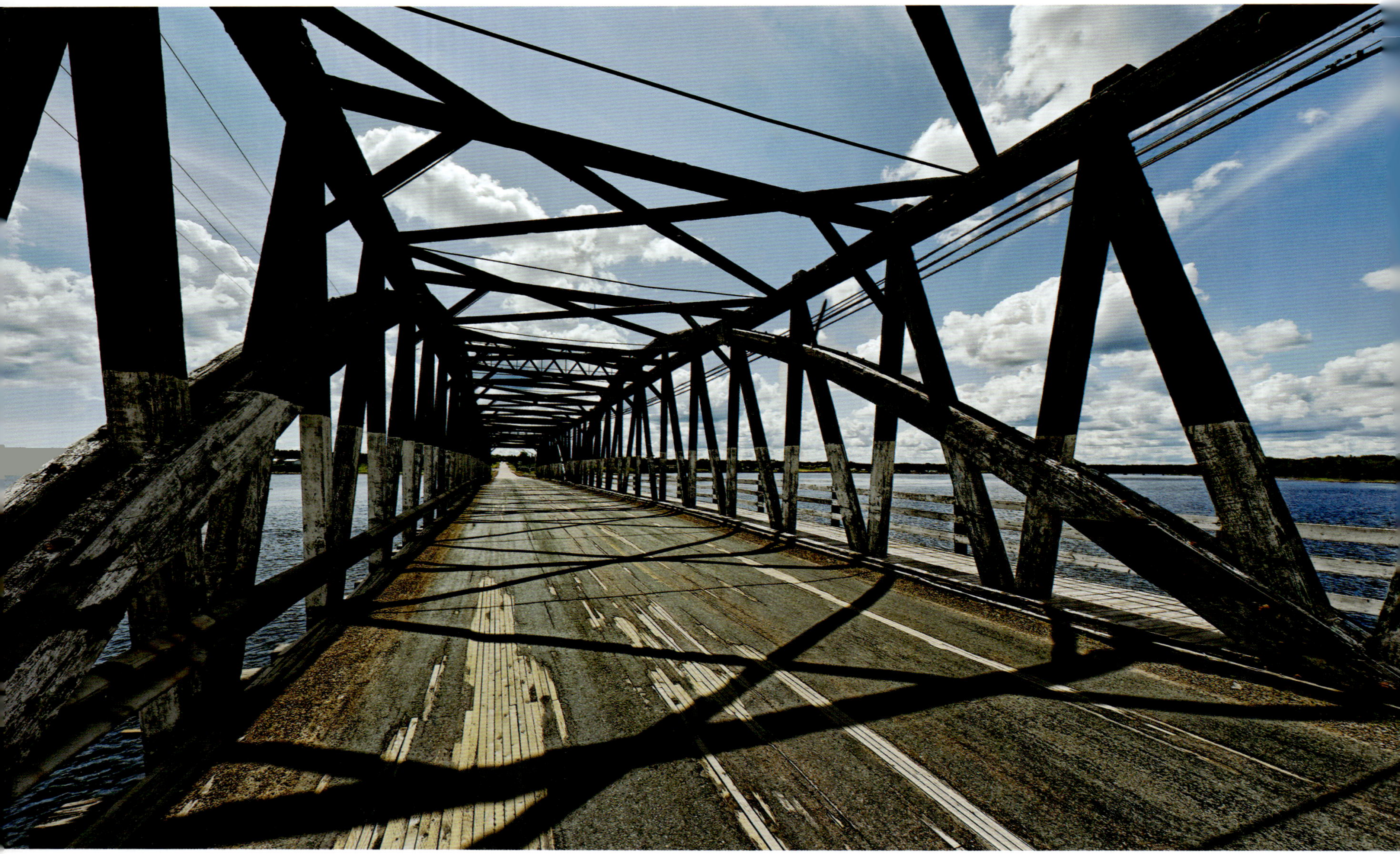

Bridge, Little Bouctouche River, near Bouctouche NB

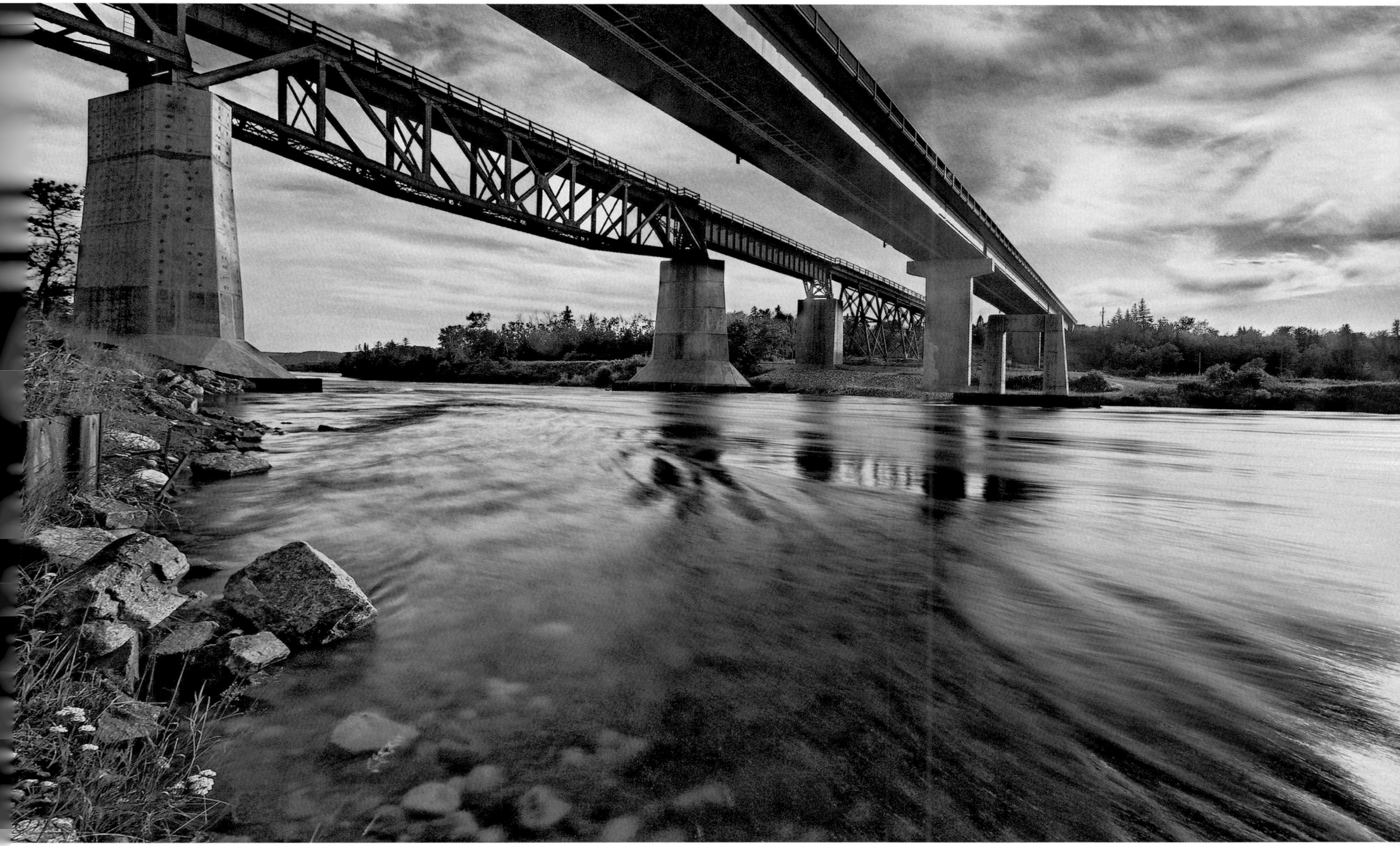

Road and Rail Bridges over Nipigon River, near Nipigon ON

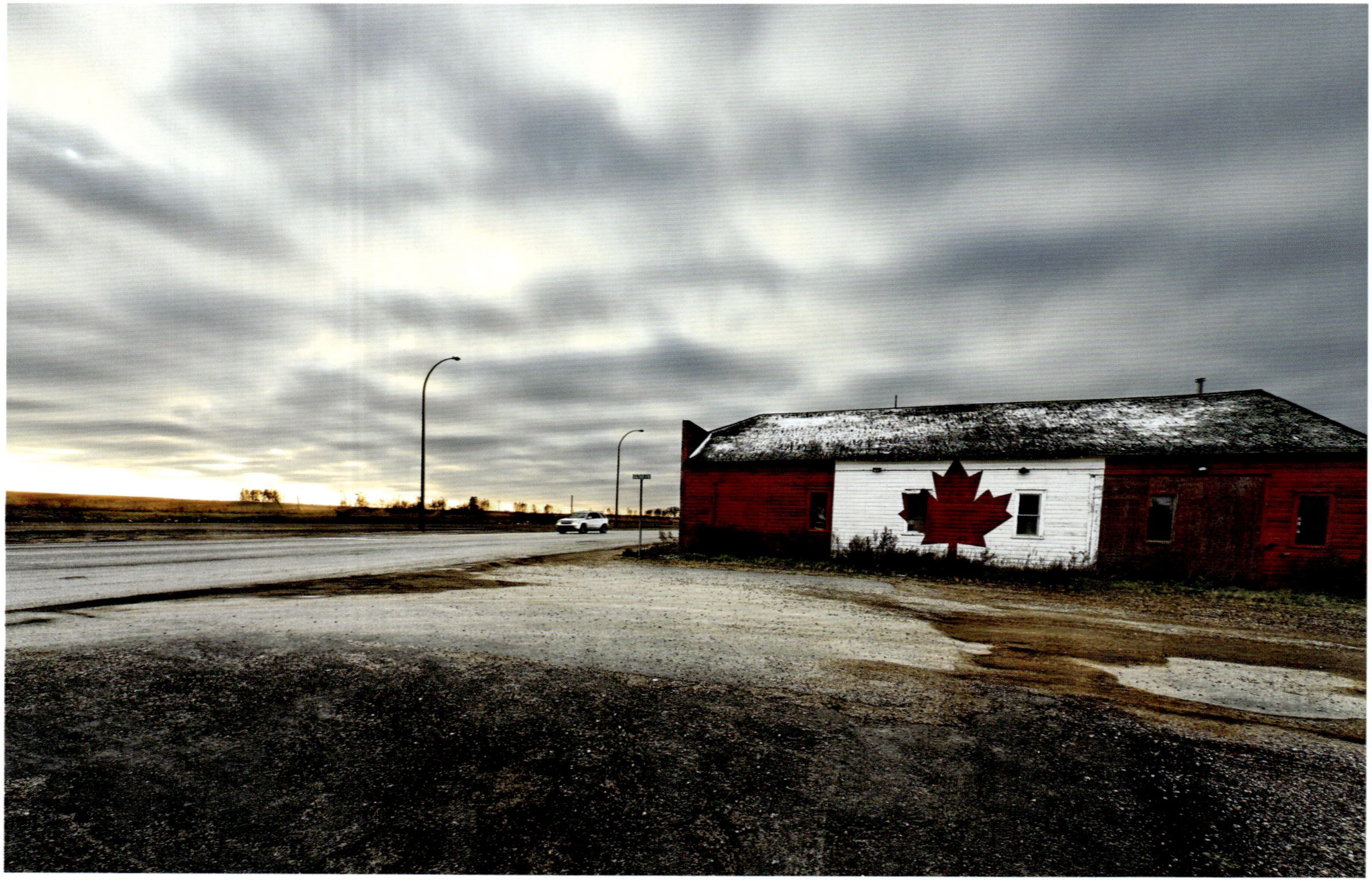

Canada, Chamberlain SK

Farm, Village, and Town

1

PARKINSON'S LAW, coined in 1955 by Cyril Parkinson, the British scholar of public administration, dictates that "work expands so as to fill the time available for its completion." Similarly, the more of a resource we have, the more of it we are likely to use, even if we don't really need to use very much.

Having observed, over the course of gathering the photographs for this book, hundreds of Canadian villages and small towns, I have concluded that a corollary of Parkinson's Law applies to streets in small towns, namely that "the more land available for streets, the wider the streets will be," or, "the smaller the town, the wider the streets." The law applies with special force to settlements on the Canadian prairies. Many times I would drive into an isolated prairie hamlet and marvel at the width of the streets while wondering why so much land had to be used to service the transportation needs of so few vehicles and people.

I am by no means the first person to make this observation. Stephen Leacock, in *Sunshine Sketches of a Small Town*, published in 1912, said of the fictional Canadian town of Mariposa:

The town, I say, has one broad street that runs up from the lake, commonly called the Main Street. There is no doubt about its width. When Mariposa was laid out there was none of that shortsightedness which is seen in the cramped dimensions of Wall Street and Piccadilly. Missinaba Street is so wide that if you were to roll Jeff Thorpe's barber shop over on its face it wouldn't reach half way across.

Leacock would have smiled wryly in contemplation of the lavishly broad streets of Beechy, Saskatchewan, population 243 (*Beechy Hotel*, p. 28), and Champion, Alberta, population 384 (*Champion Inn*, p. 129). (I should point out that the photographs in Beechy and Champion were shot with a wide-angle lens that exaggerate to some extent the apparent width of the streets. Nevertheless, the streets are indeed very wide!)

2

Embedded in small-town roads are countless stories of arrival and departure. The quiet roads and peaceful farmland around Grand-Pré, Nova Scotia (*Grand-Pré*, p. 103) are today mute witness to a sorrowful story of departure—the expulsion of the Acadians from the Annapolis Valley in the middle of the eighteenth century. The Acadians were descendants of French-speaking colonists who settled in what later became the provinces of Nova Scotia, New Brunswick, and Prince Edward Island. Grand-Pré was founded by Acadians in the late 1600s. Through the construction of a network of dykes to hold back the tides of the Minas Basin, the settlers developed great expanses of what had been tidal salt marshes into productive farmland. But the Acadians fell victim to the battle between the colonial powers of France and Great Britain for supremacy in North America. Distrustful of the approximately 23,000 Acadians living in their territory, the local British authorities ordered them expelled to other French and English colonies or to France. Many thousands died as a result of the upheaval; according to some records, only 10,000 survived the deportation. In 2003 Queen Elizabeth II of Great Britain issued a proclamation acknowledging Britain's responsibility for the forced removal and the "tragic consequences" of the expulsion.

A contrasting story—one of refuge and prosperity—unfolded about a century later in the similarly flat and fertile land of the Red River valley in southern Manitoba. Many German-speaking Mennonites settled in the area. They came to southern Manitoba from the Russian Ukraine in the 1870s to escape religious persecution. They were expert grain famers and so were attracted to the rich farmland of the Red River valley. In the Mennonite town of Altona (*Storefront*, p. 118), a local man struck up a conversation with me, and was interested to learn that my mother's own history had important similarities to the story of the Mennonite immigrants. My mother's story, too, was one of escape from persecution, and it likewise brought her and her family to southern Manitoba. They settled in nearby Winkler in the 1930s after emigrating from Poland to flee the rising anti-Semitism that followed the ascent to power in Germany of Adolph Hitler. My mother, who arrived in Canada at the age of four, had enduring memories for the rest of her life of the Mennonite communities of southern Manitoba.

3

One predictable consequence of including roads in photographs is that I often found myself framing a shot while standing on or in the middle of the road. On large and busy roads in major cities (*McKnight Boulevard*, in Calgary, p. 180), this could be hazardous, if not downright foolish. In small towns, though, I benefited from the courtesy of residents who seemed happy to oblige my need to interfere with traffic. Motorists would wait patiently while I stood in an intersection fine-tuning a composition. Even when I was shooting from the sidewalk, I would sometimes see that drivers were stopping outside what they supposed was my field of vision, so as not to spoil my shot! Cyclists and pedestrians, too, would get into the act, hesitating to walk in front of me. "Go ahead. Don't worry about me," I would say, and wave them on. I did this not only out of a desire not to inconvenience anyone but also because there were occasions when I felt that having a pedestrian or cyclist in the frame was important to the photograph (*Queen's Hotel*, p. 119).

As well, people in villages and small towns were much more likely than in larger towns and cities to approach me to ask what I was taking photographs of and why. (Indeed, I can't remember this having happened once in a larger place.) While most of the encounters were friendly—as with the man in Altona—not all of them were. People sometimes are suspicious of photographers, especially when they see them shooting things that may not appear to be natural subjects for photographs. And a photographer on a road in the middle of farm country or in the main street of a lonely village tends to invite attention.

For example, not far from Altona near Morris, Manitoba (*Morris*, p. 121), I stopped my car to photograph a field of burning wheat straw (*Fields on Fire*, p. 57). A man driving an ATV through the field, igniting piles of straw with a propane torch, pulled up beside me to ask if I was "from an activist group." Farmers are sensitive to public concerns about air pollution caused by burning straw after the harvest. Most grain farmers in Manitoba don't burn their wheat straw, but when the soil has a high clay content, and in years when yields are high, it becomes too difficult to till the straw back into the ground. That same evening, when I went for a walk in Winnipeg, about sixty kilometres north, the air was heavy with the smell of the burning fields.

In Tiverton, Ontario, as I stood in the middle of the main street, a young man stopped his car next to me, informed me that I was photographing his "wife's building" and asked, "Is everything all right?" I told him that he had nothing to worry about, but he drove off muttering that my interest in the building was "very strange."

Other encounters were oddly amusing. In Wallaceburg, Ontario, the red brick building in *Cyclist* (p. 184) happened to be for sale. The real estate agent who had the listing was, by coincidence, walking with friends down the street at the moment when I was pointing my camera in the general direction of the property. "Hello! Hello!" I heard, and then looked up to see her waving her arms to get my attention. "Are you interested in the building?" she asked.

In Black Rock on the Fundy coast of Nova Scotia (*Puddle*, p. 82), I met a man walking on the beach with his grandson. He asked me if I had spotted the sea monster he said was often observed near there. Playing along—I thought he was trying to amuse the boy—I said I hadn't seen it yet. But a few more moments of conversation made it clear to me that he was serious about the "monster." He asked for my email address and later sent me what was supposed to be a photograph of the beast.

4

About three-quarters of Canadians identify themselves as Christian. It follows that as one travels around the country one will never be far from a church or a Christian cemetery. And churches and cemeteries will, by their nature, always be located on roads.

The implications for me of these simple and unremarkable facts were significant. Churches and cemeteries were often in view as I travelled the country taking photographs of roads. With churches and cemeteries go public displays of symbols that are integral to the Christian faith—in particular, the cross. Until I embarked on this photographic project, it had never occurred to me just how omnipresent Christian symbols are in our public spaces.

Perhaps it was the effect of repetition; perhaps it was my susceptibility to the simple, austere geometry of the Latin cross; but there were times when I had to restrain myself from taking yet one more photograph of a church or a cemetery. Though neither a religious person nor a Christian, I found myself impressed and moved by the visual and emotional power of Christian symbols in the landscape. To my eye—and I believe the photographs reflect this—that power is much greater in rural areas and small towns than in cities. In a city the visual impact of a church is easily drowned out, and human weakness is masked by the seeming solidity of office towers and the roar of traffic. In rural settings, where the buildings are smaller, the sky broader, and the roads emptier, humans look more vulnerable, alone, and small; more in need of the divine sustenance that only a church can promise. How better to express this than through a photograph of the Spring Creek Lutheran Church (*Spring Creek Lutheran Church*, p. 109) near the village of Hanley in south-central Saskatchewan? The tiny building with its fragile-looking spire stands alone on the prairie beside a narrow gravel road. I saw a similar message—"look up and you shall be saved"—in the juxtaposition of the seemingly upward-pointing arrow and the tiny cross of the Assumption Roman Catholic Church in Avondale, Newfoundland (*Arrow and Cross*, p. 113). And indeed, the idea of looking up and being saved has certainly proven true, at least in a secular sense, where this church is concerned. The building sits high atop a hill, and fishermen of old making for Avondale harbour traditionally used its steeple as a navigational aid.

Neon Cross (p. 112) was perhaps the most irresistible example, for me, of the power of Christian roadside imagery. I was driving swiftly (rather too swiftly, given the number of moose near the highway on a summer evening) down Highway 8 near Blackville in central New Brunswick. It was late in the evening in mid-July, and I still had about ninety minutes to go before reaching my hotel in Fredericton. I had been driving and taking photographs since 10:00 that morning. The light was fading, my eyes were tired, and I felt I was at last done for the day. But when the white clapboard church came into view with the illuminated cross atop the spire backlit by the last rays of the sun, the image looked irresistible to me. I pulled over onto the shoulder, set up my tripod, and took a series of long exposures—as many as I could bear before I had to retreat to my car to escape dusk's squadron of mosquitoes.

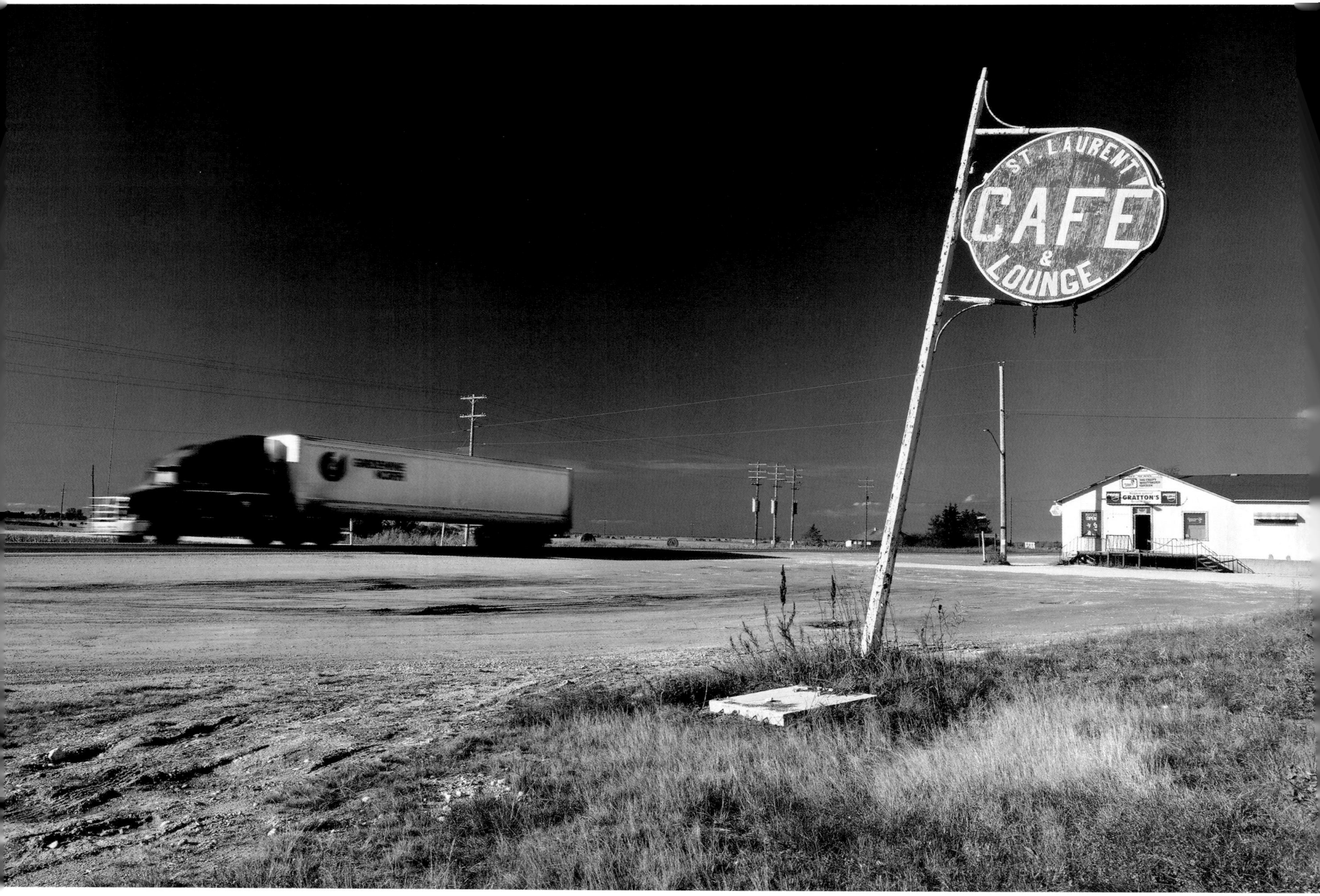

Sign Board, St. Laurent MB

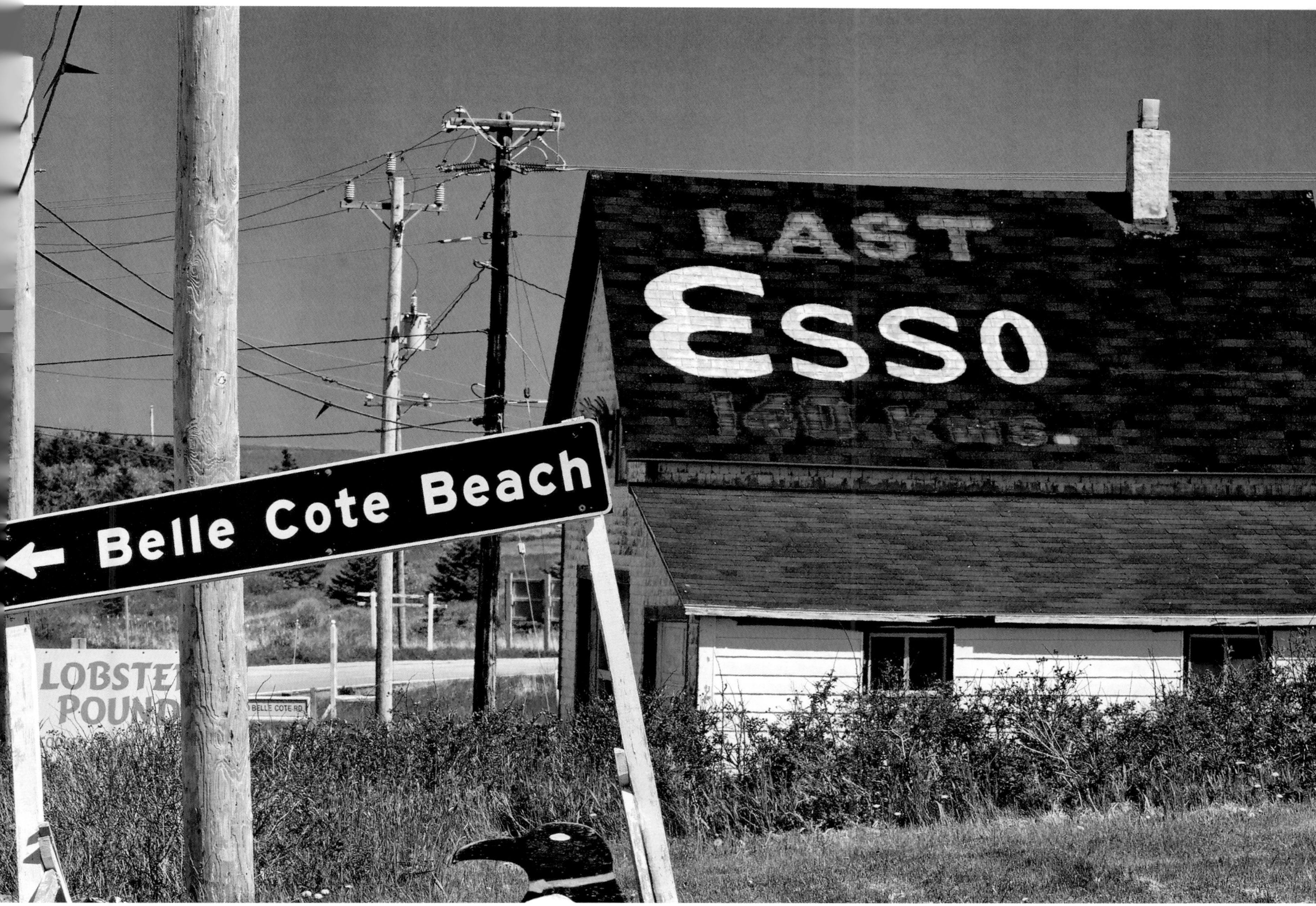

Last Esso, Belle Cote NS

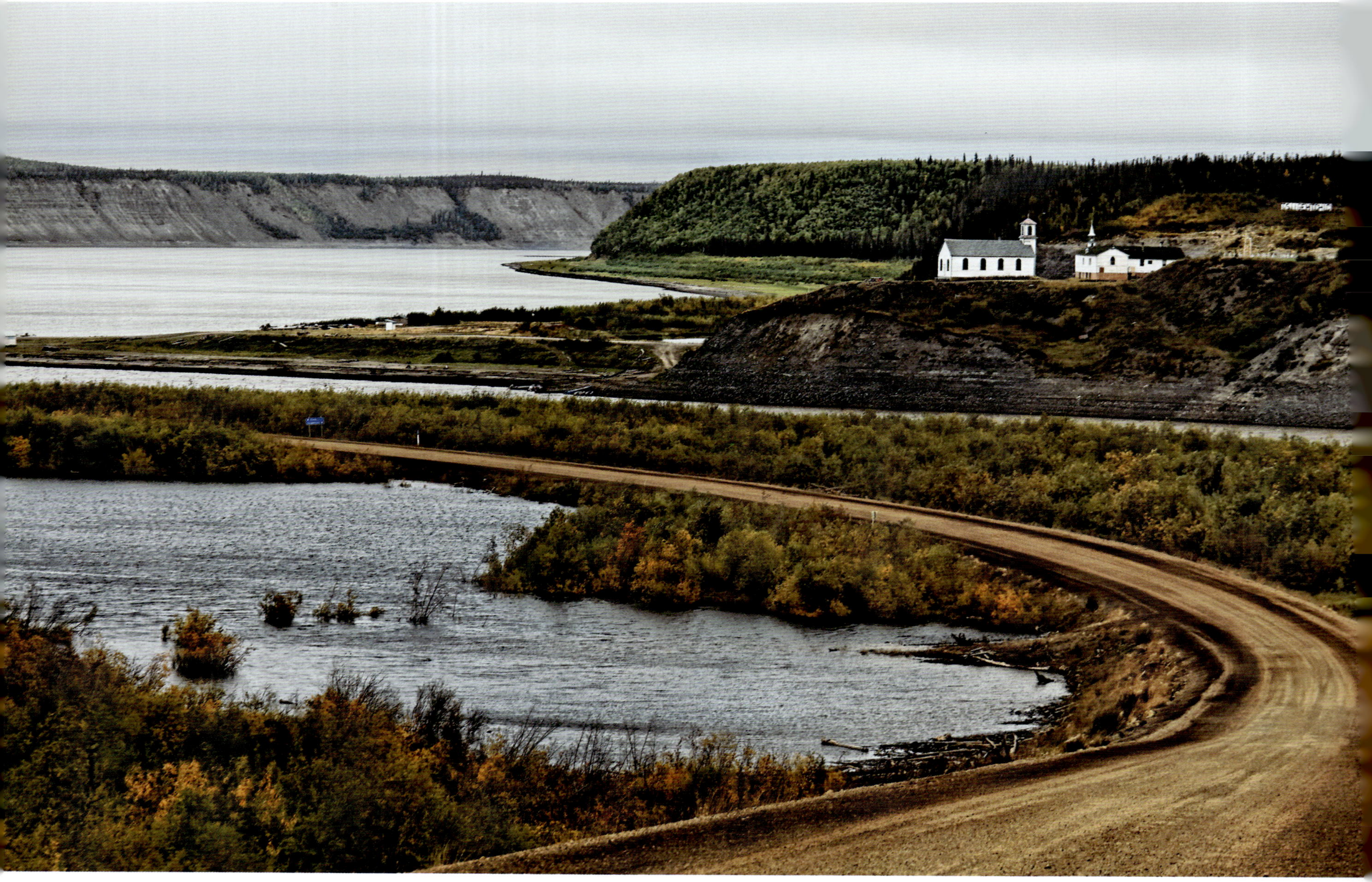

Tsigehtchic, Tsigehtchic NT

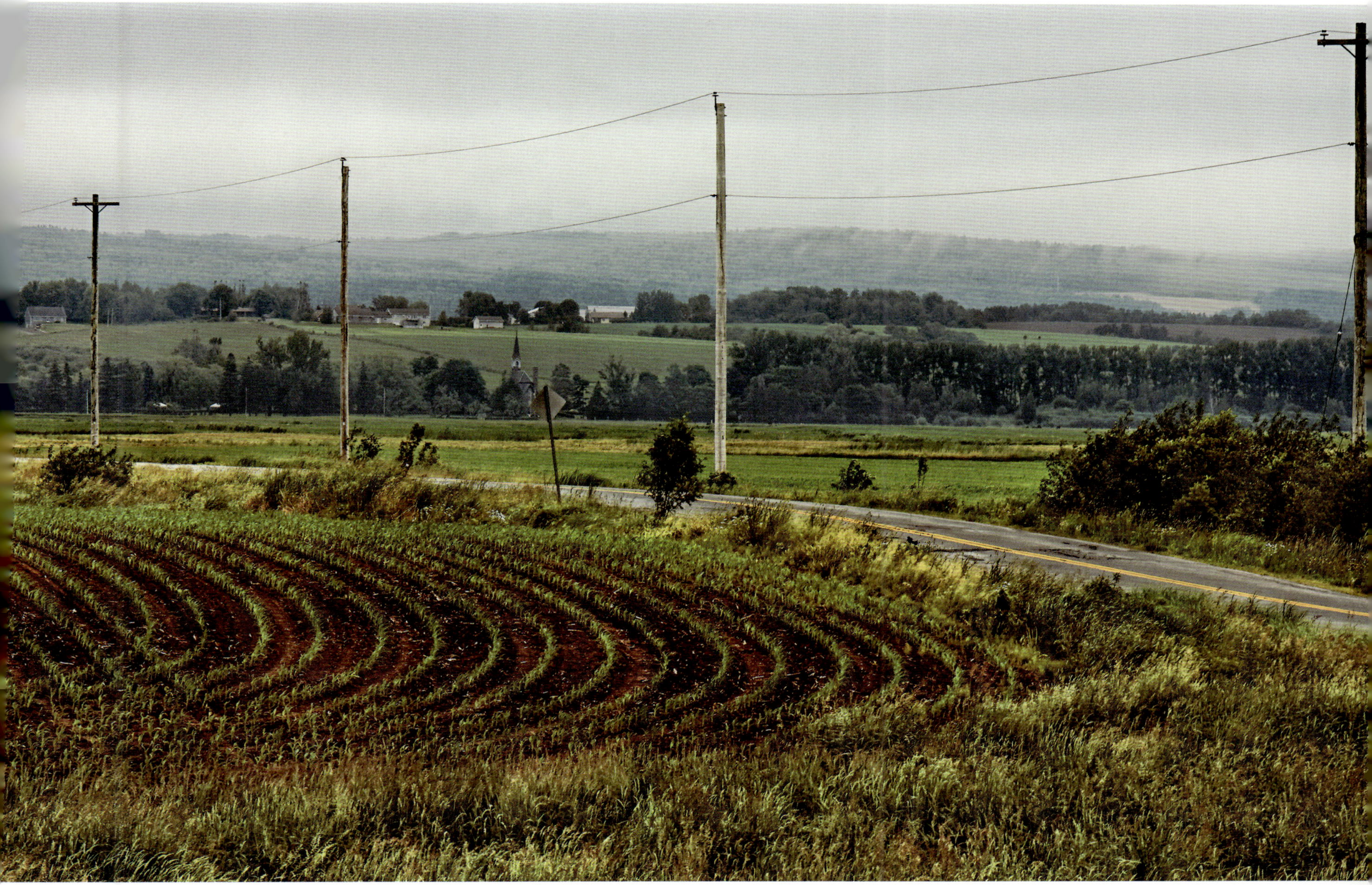

Grand-Pré, Grand-Pré NS

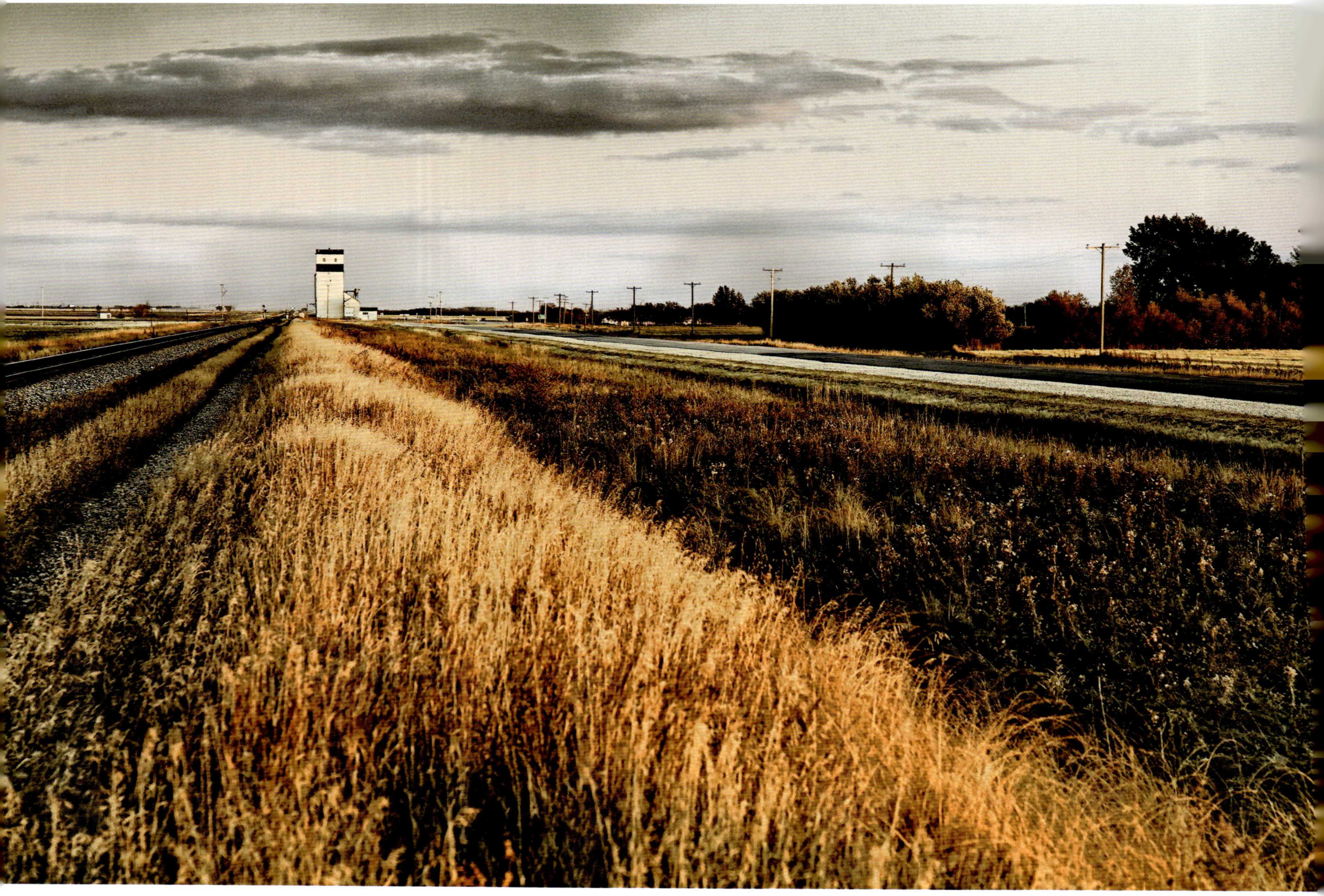

Sunset, Meadows MB

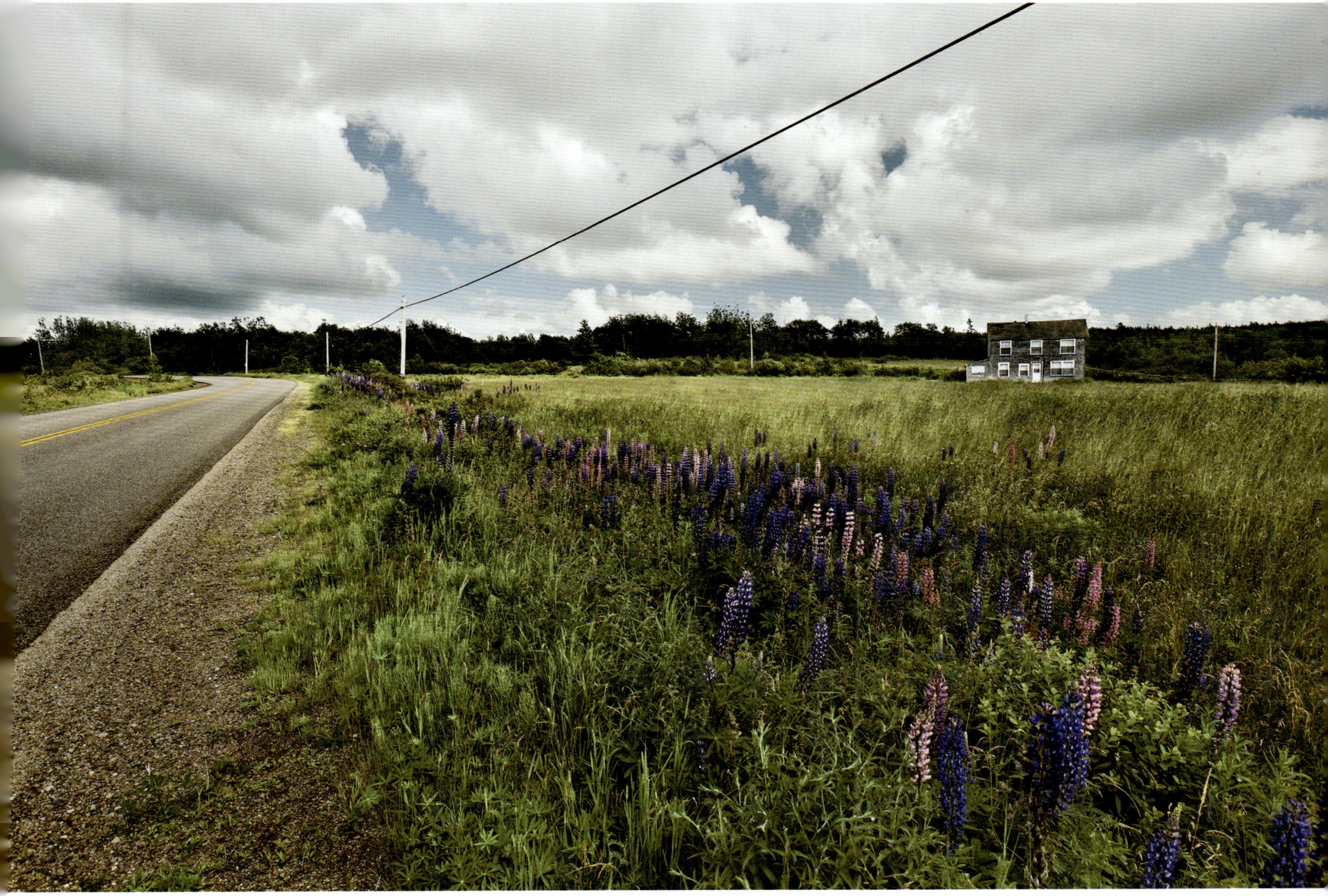

House and Lupines, Phinneys Cove NS

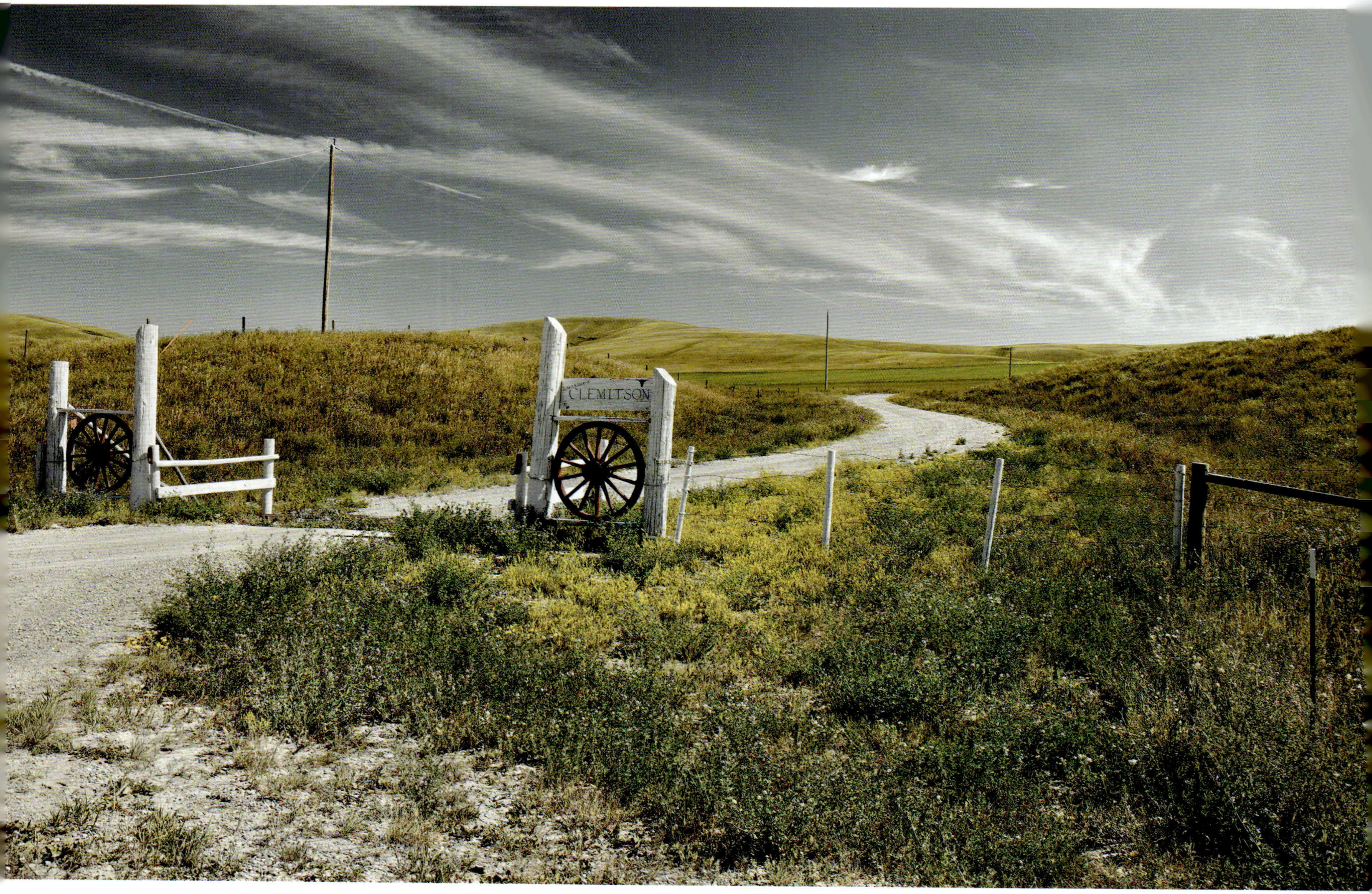

Gate, Spring Point AB

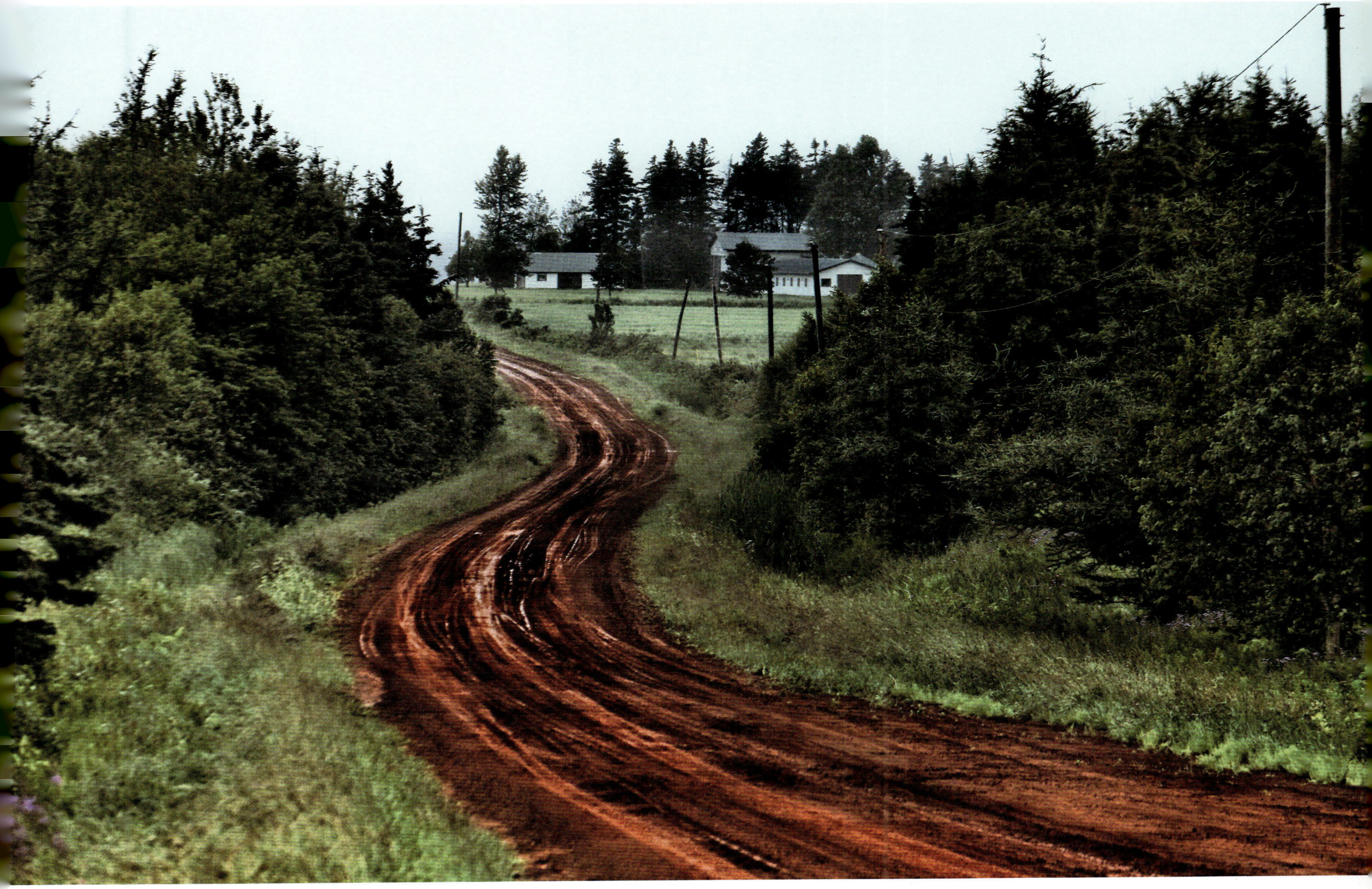

Brae Harbour, Brae Harbour PEI

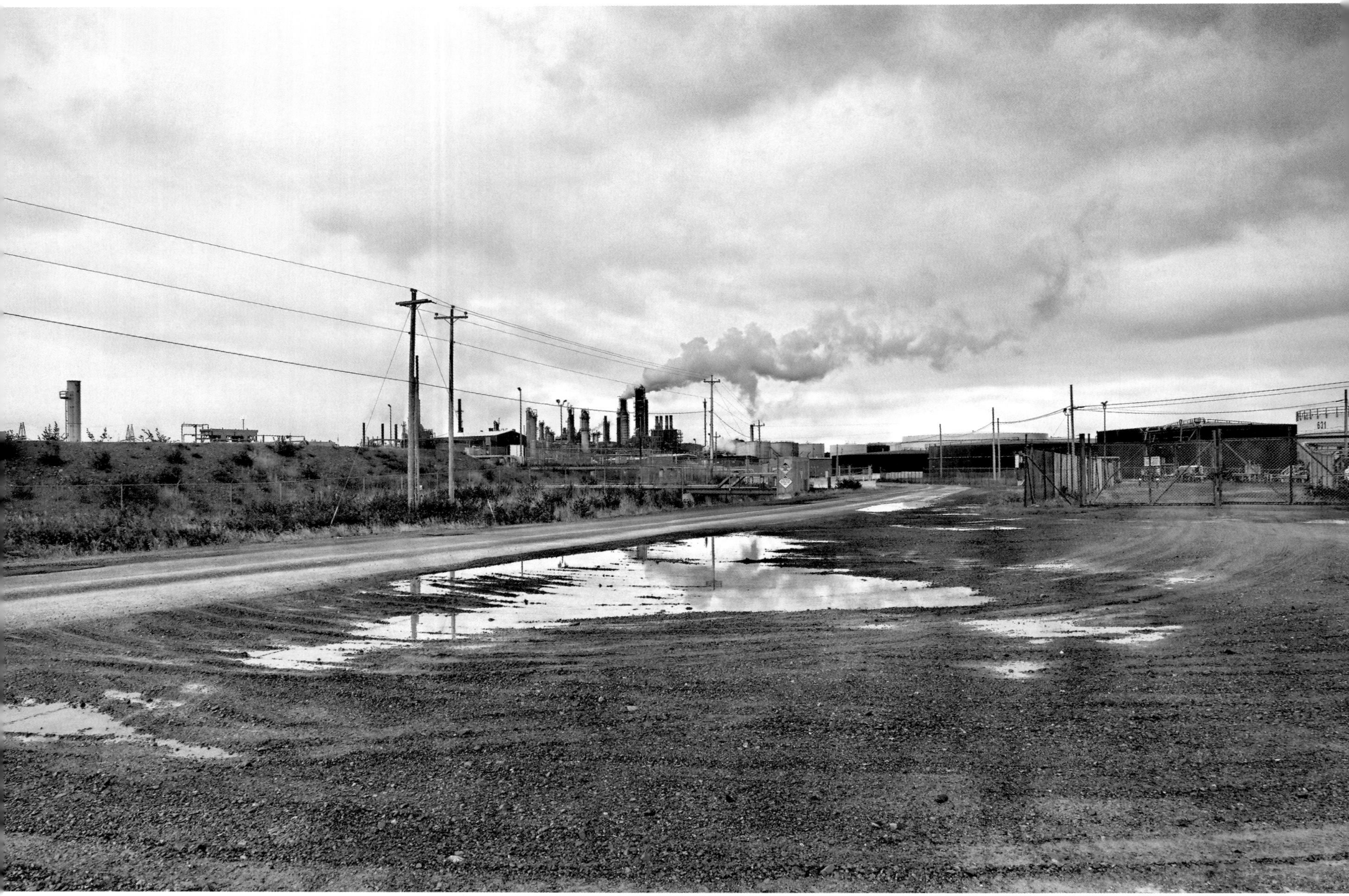

Refinery, Come-by-Chance NL

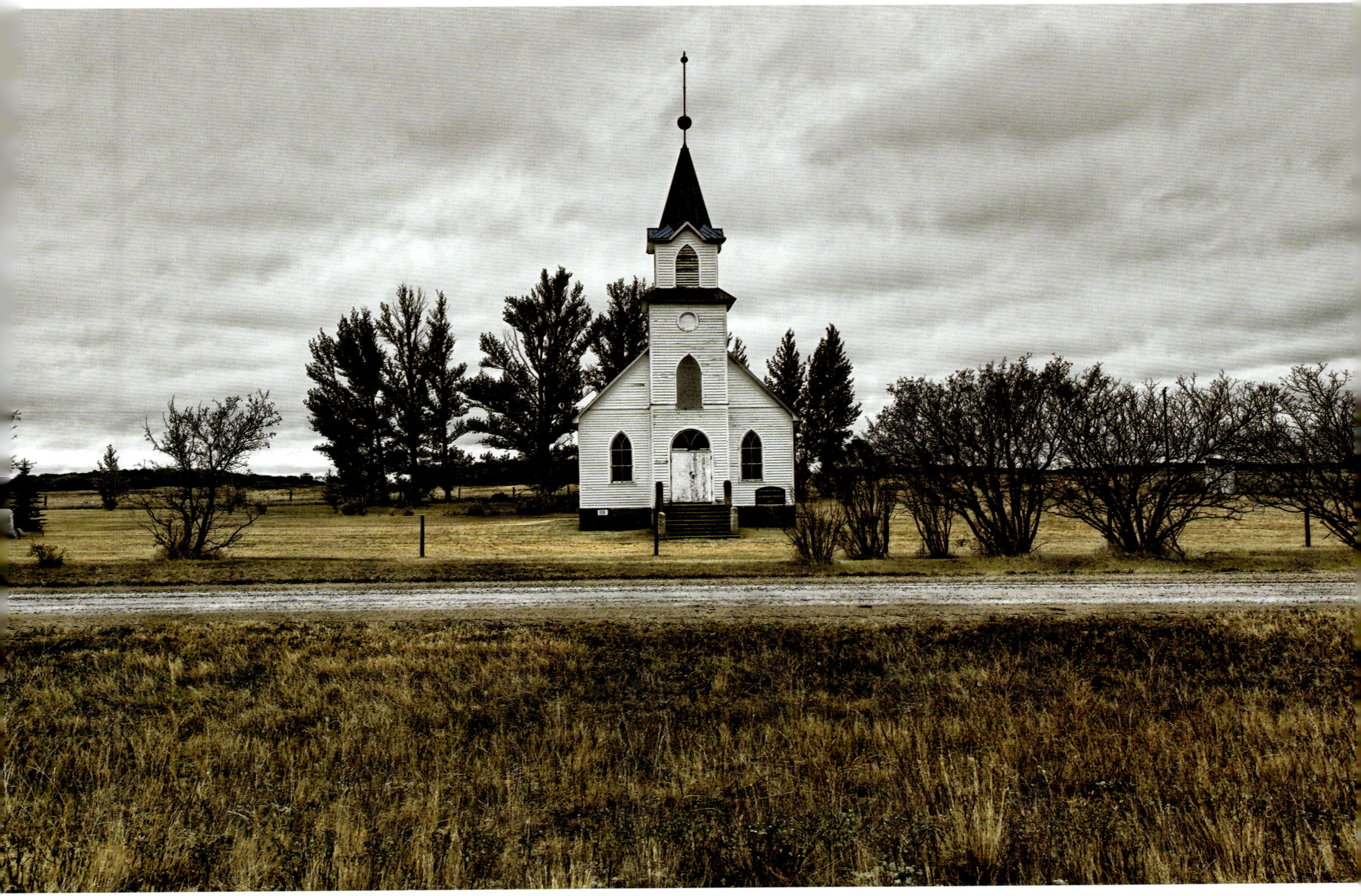

Spring Creek Lutheran Church, near Whitecap Dakota Indian Reserve SK

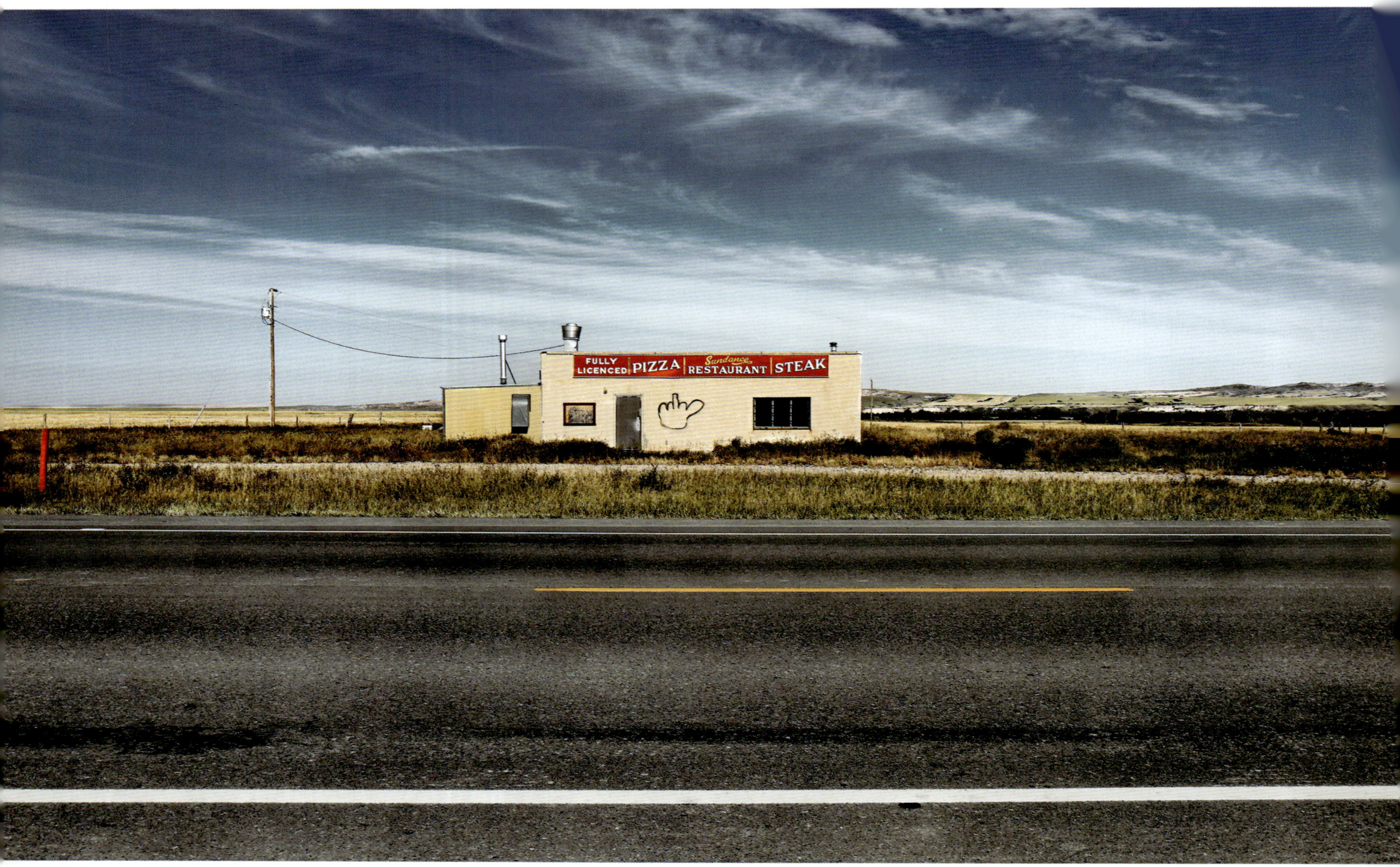

Sundance Restaurant, near Stand Off AB

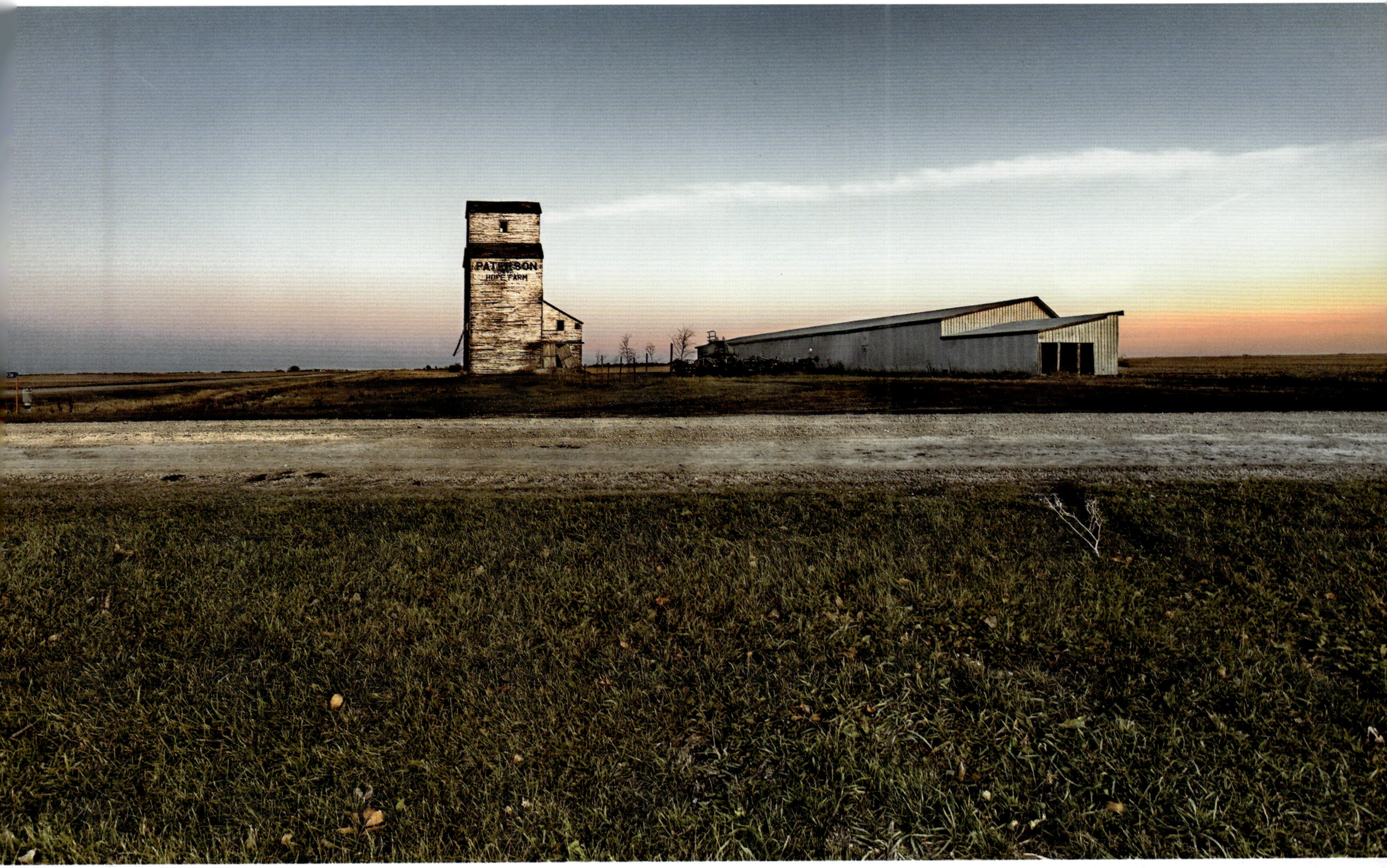

Prairie Buildings, near St. Jean Baptiste MB

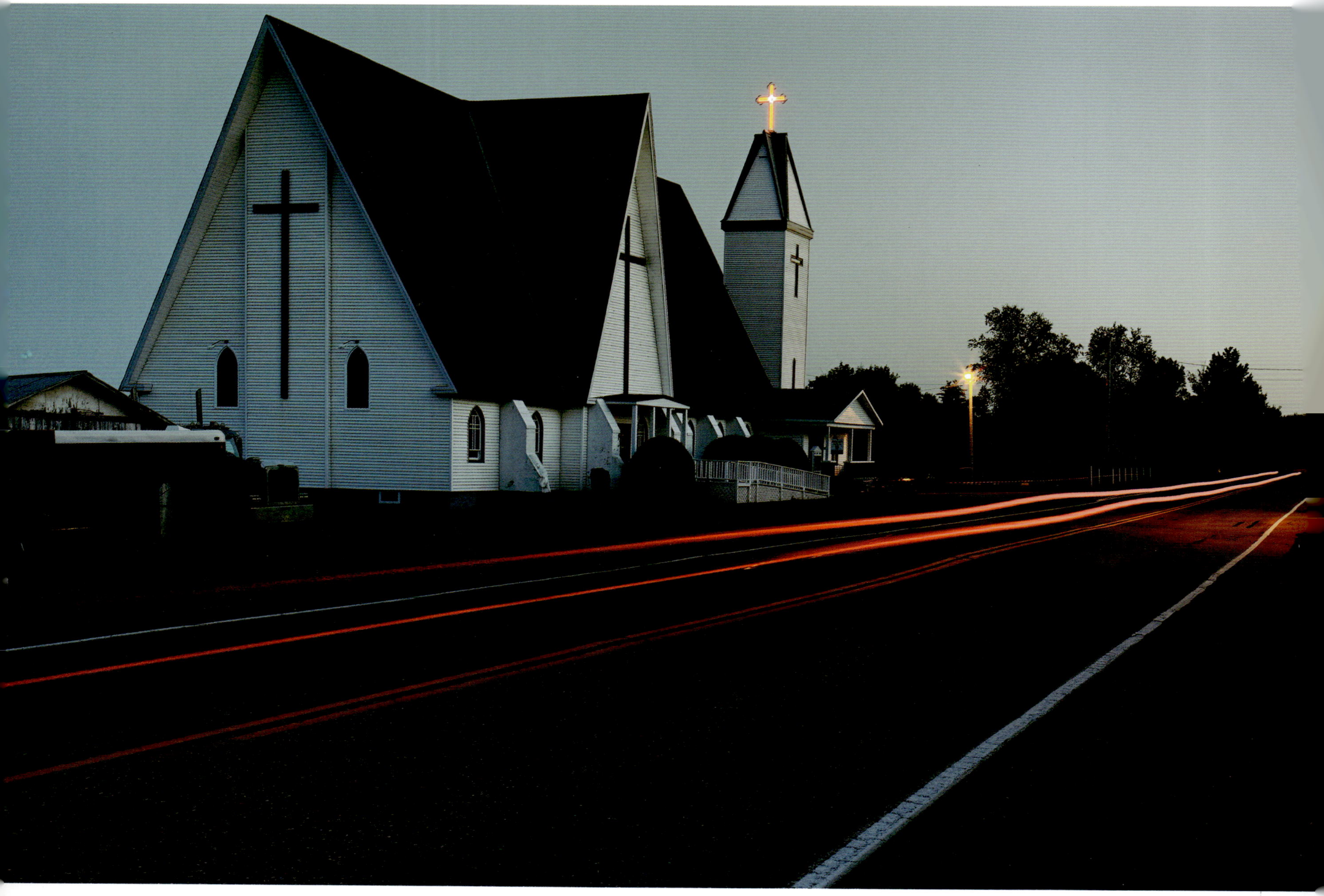

Neon Cross, Blackville NB

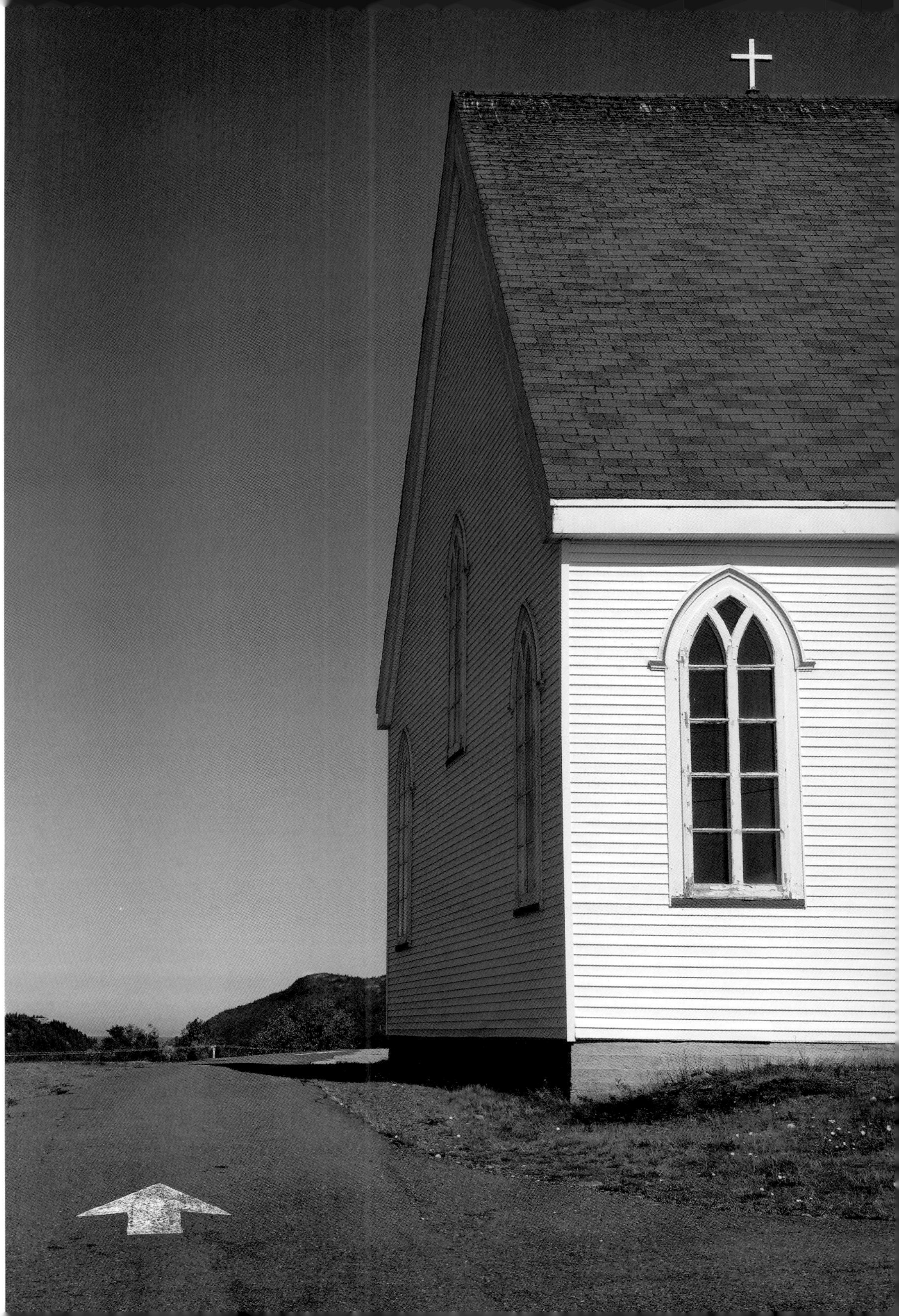

Arrow and Cross, Avondale NL

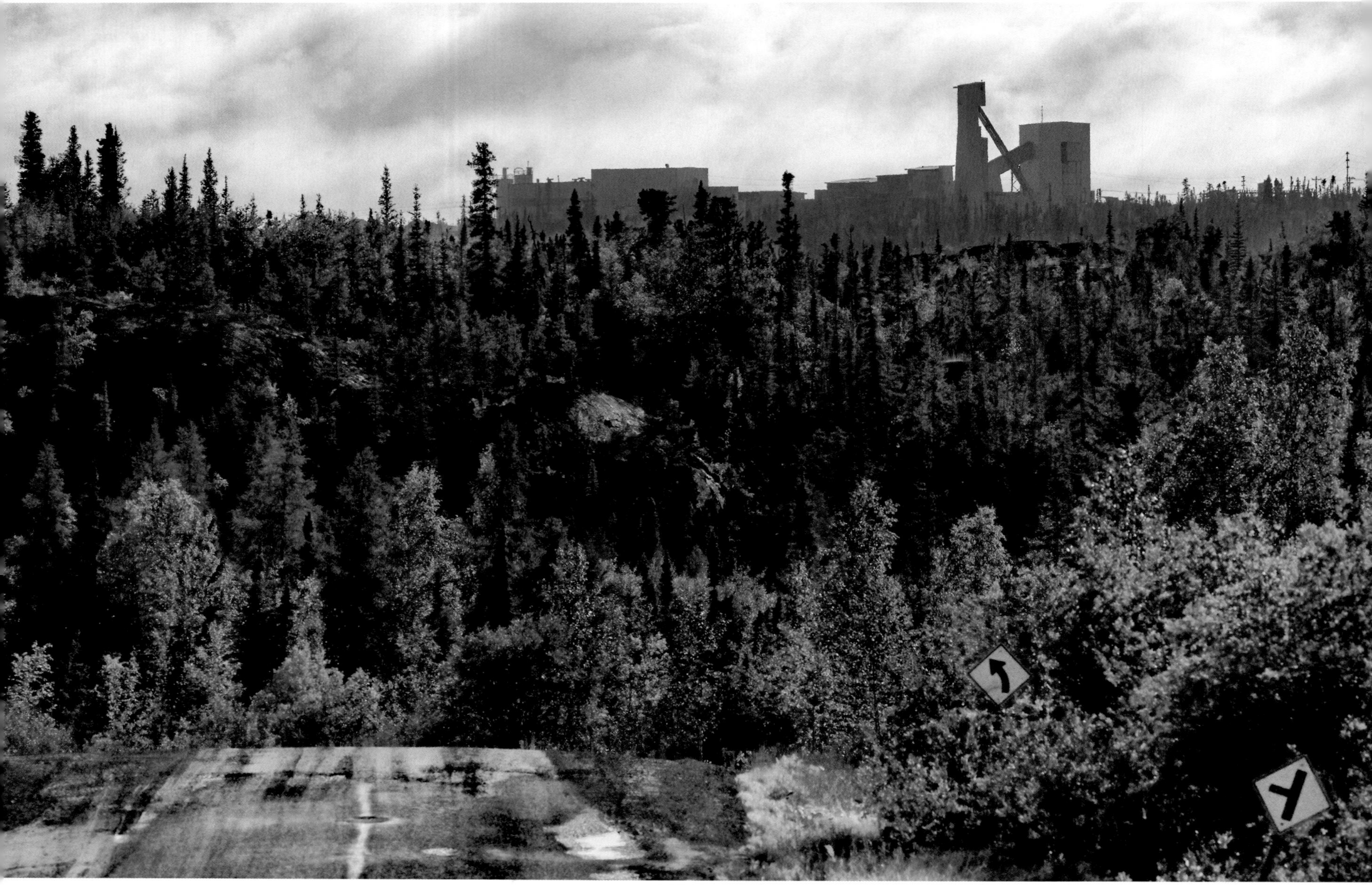

Giant Mine, near Dettah NT

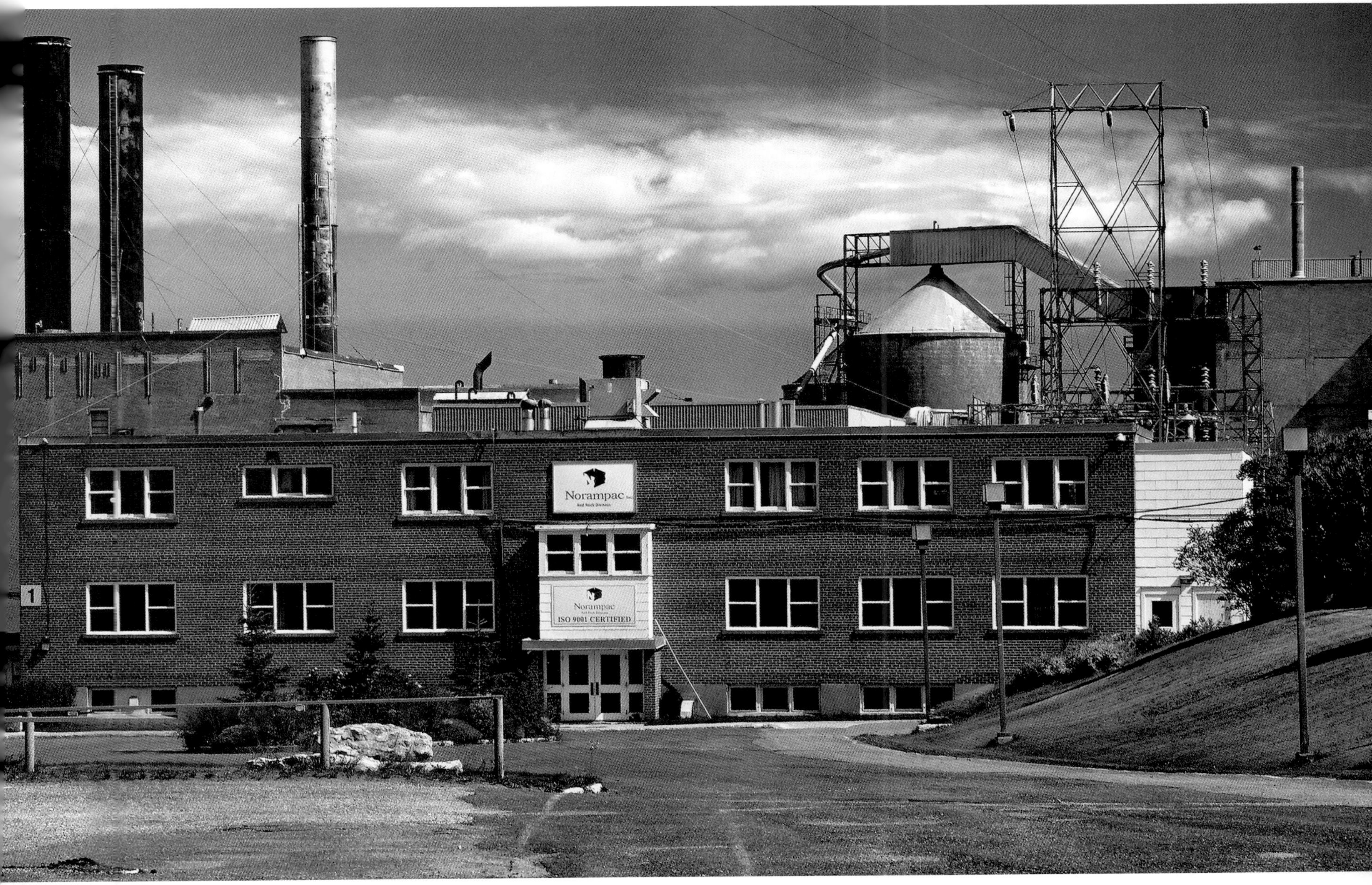

Paper Mill-Shut Down, Red Rock ON

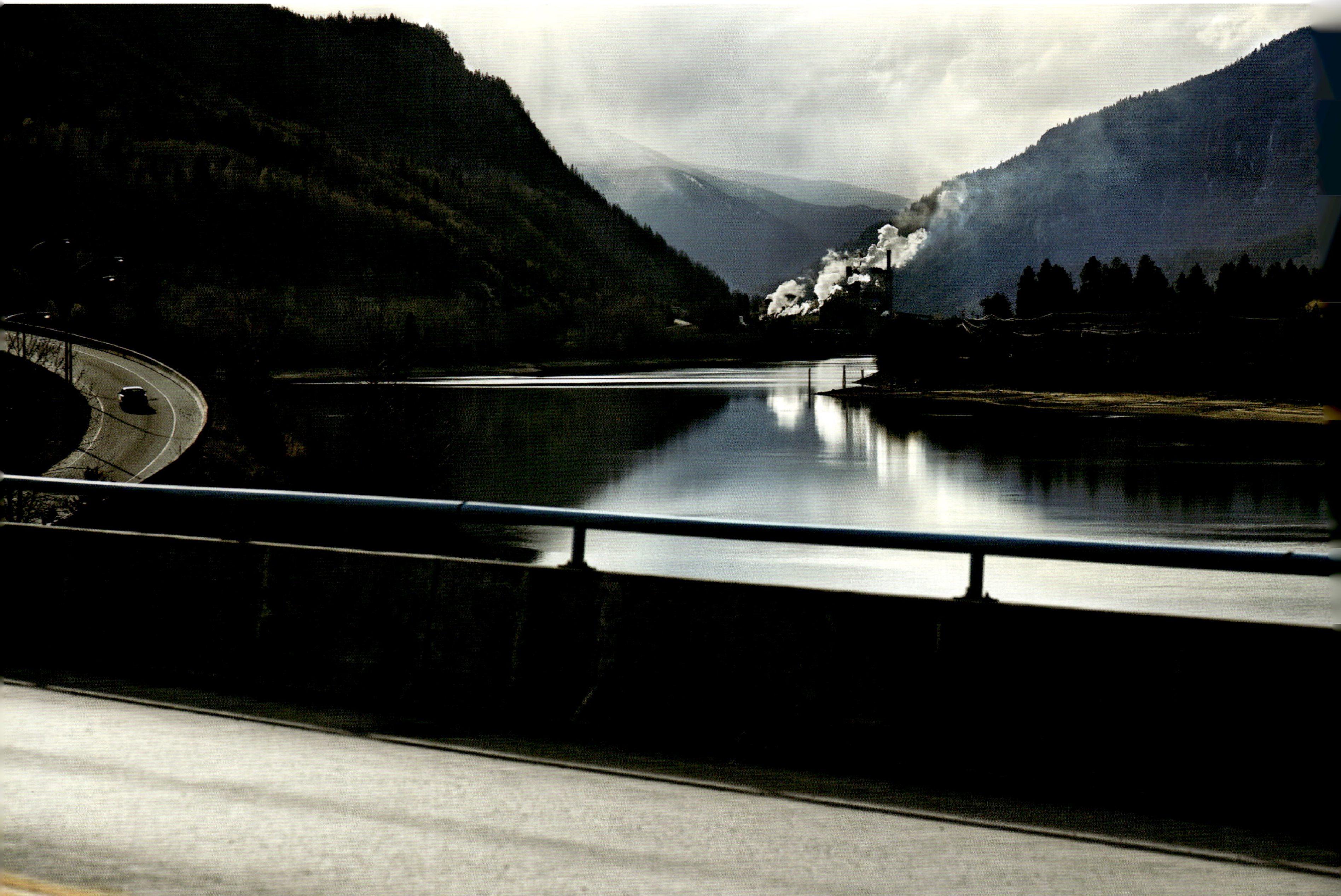

Pulp Mill, near Castlegar BC

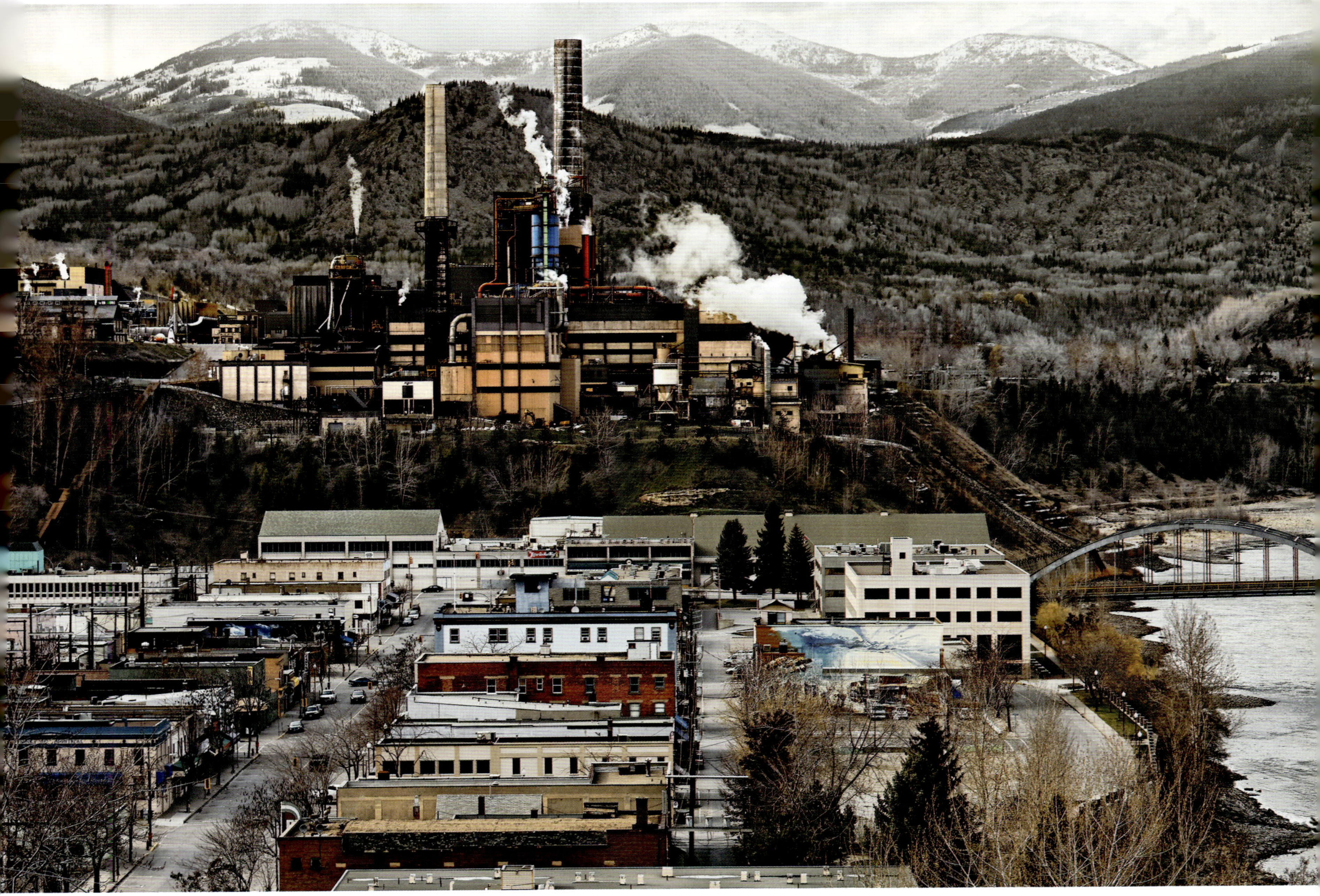

Smelter, Trail BC

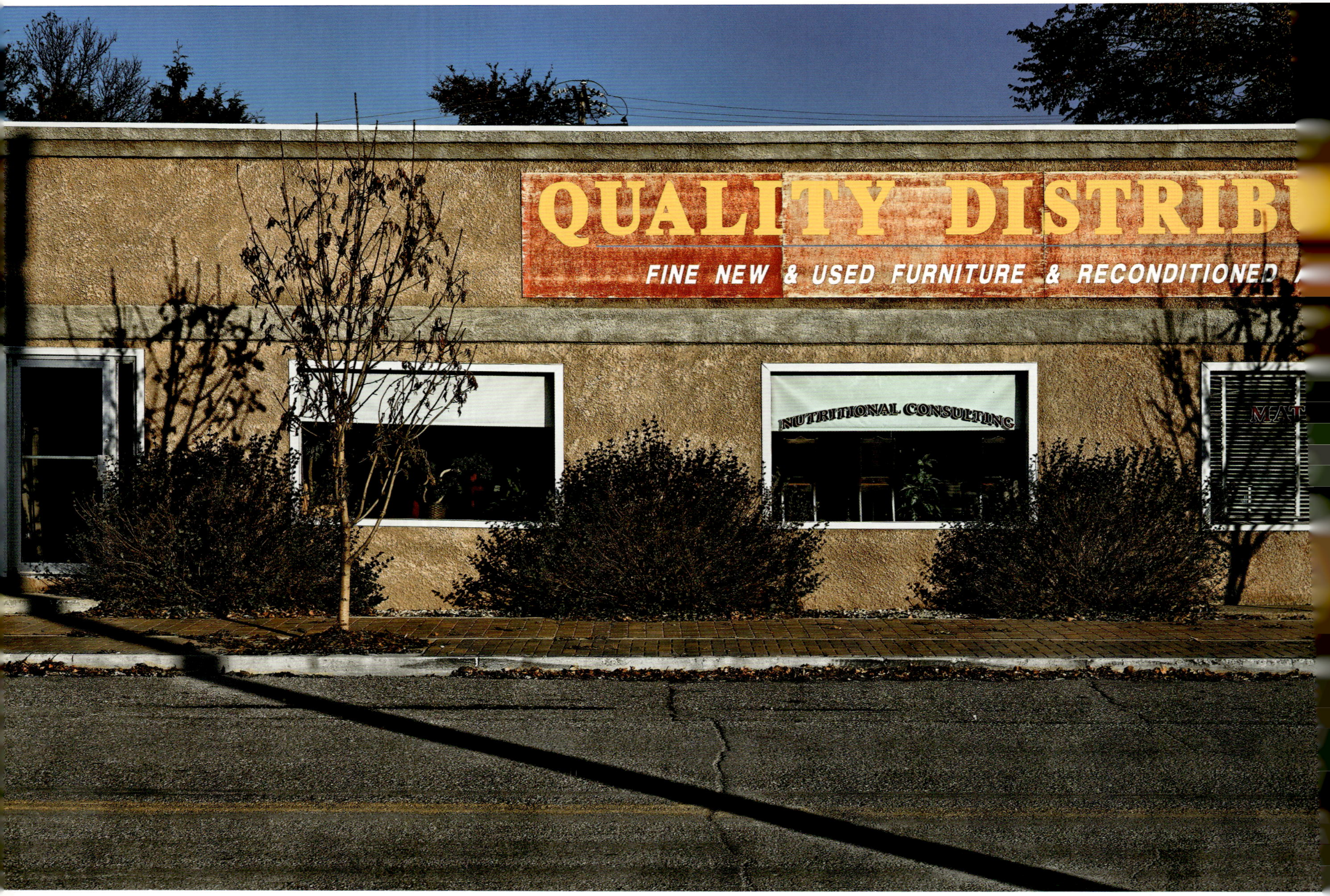

Storefront, Altona MB

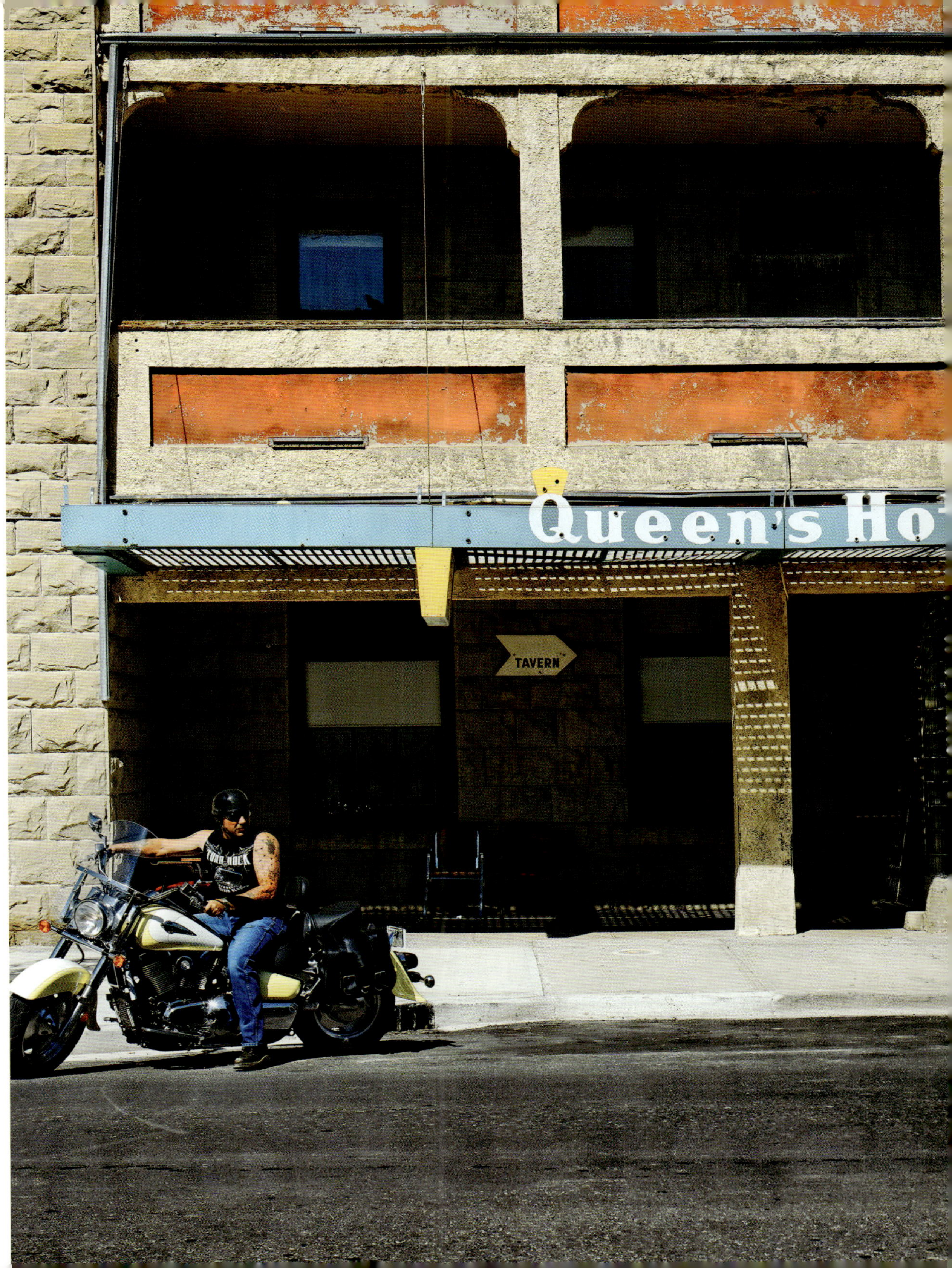

Queen's Hotel, Fort MacLeod AB

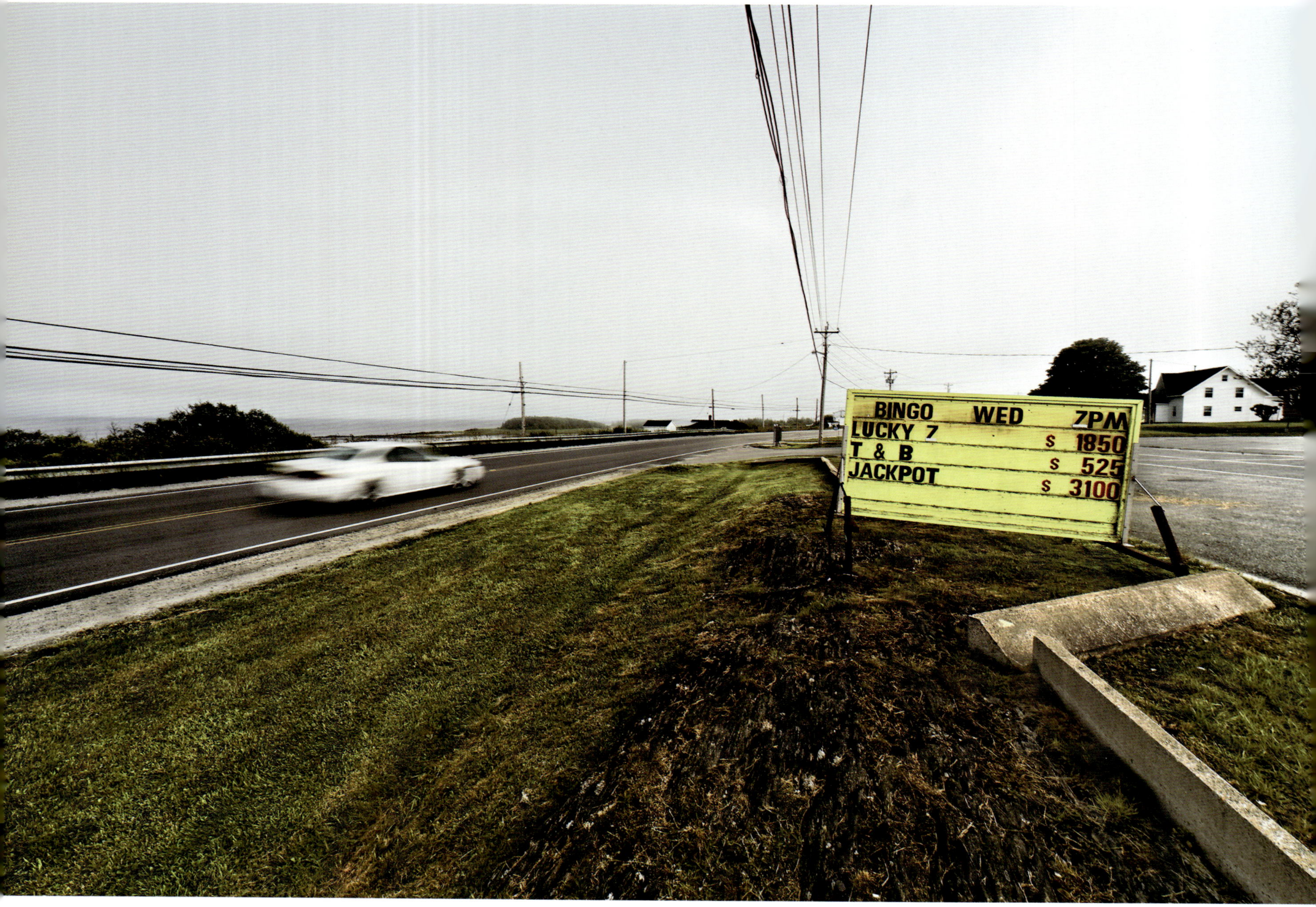

Bingo, West Pubnico NS

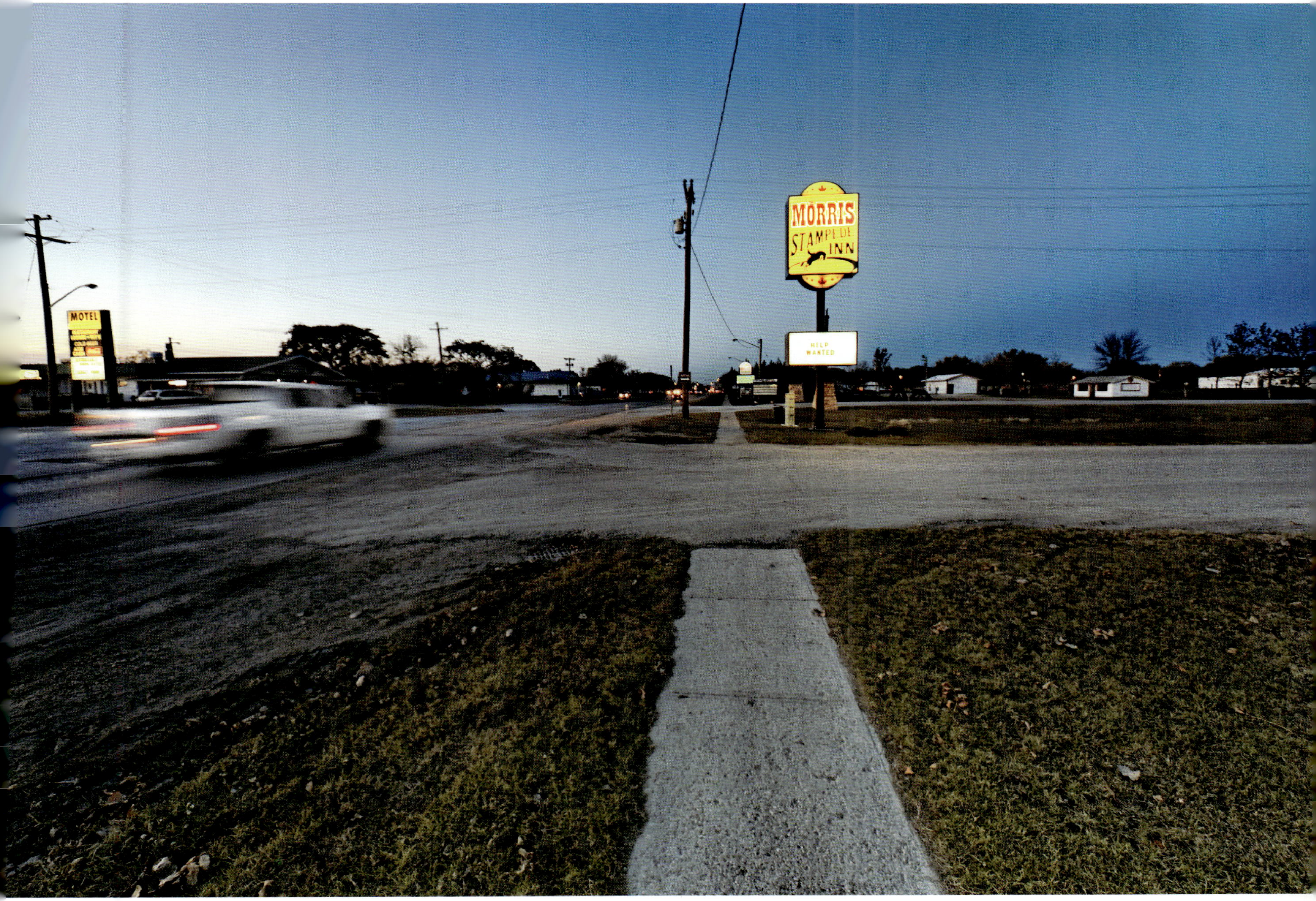

Morris, Morris MB

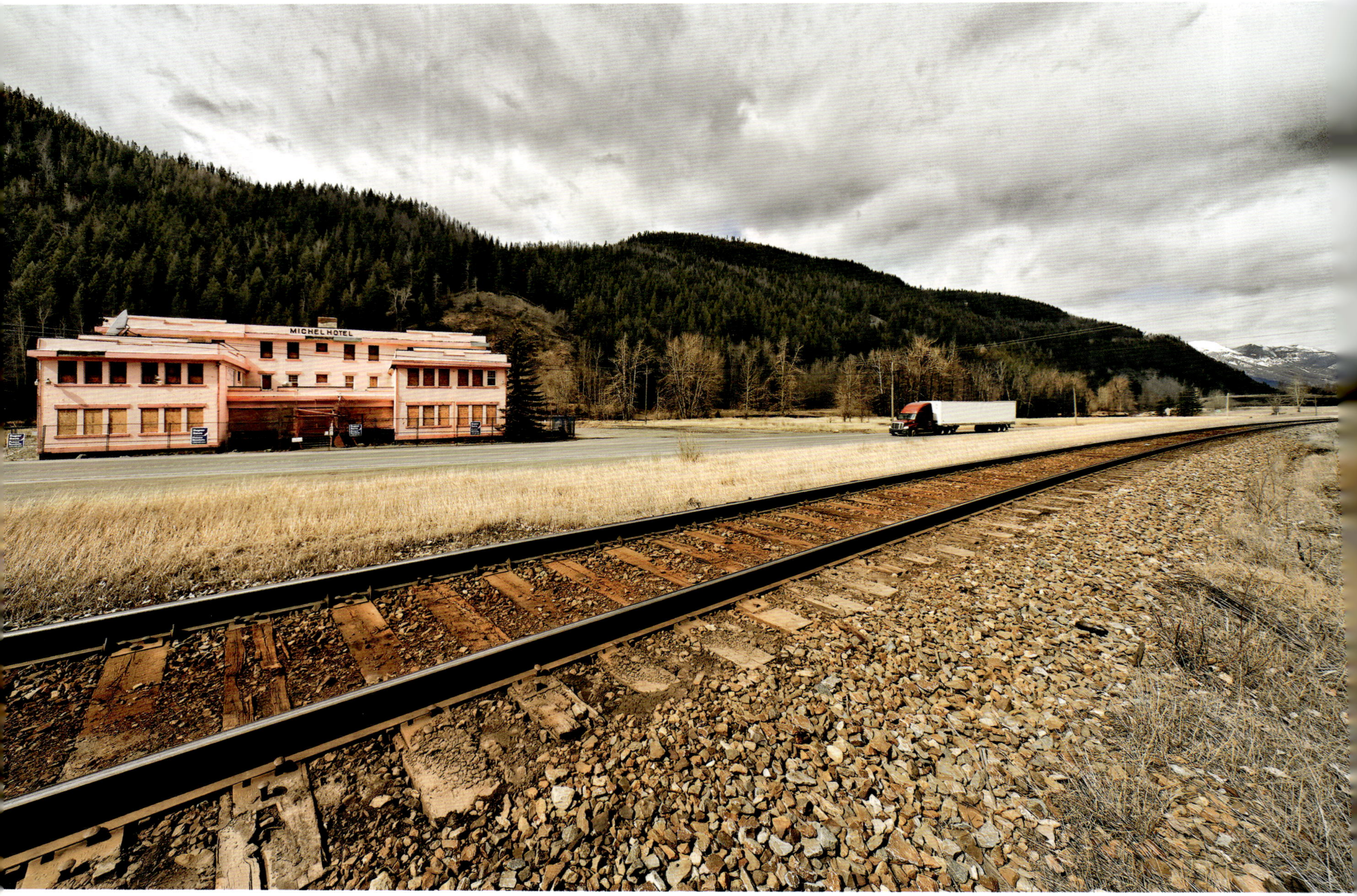

Michel Hotel, Michel BC

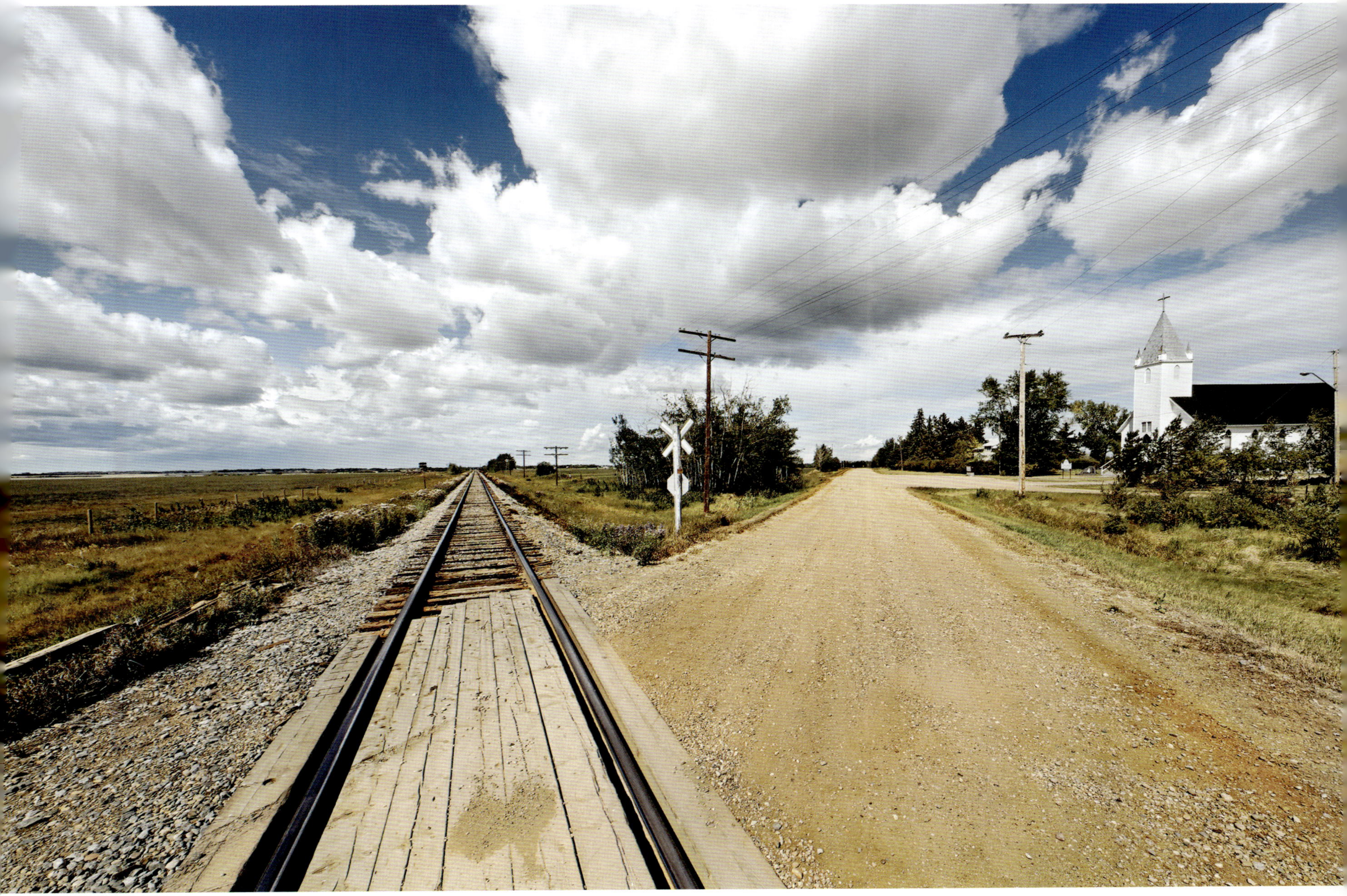

St. Charles Borromeo Roman Catholic Church, Mearns AB

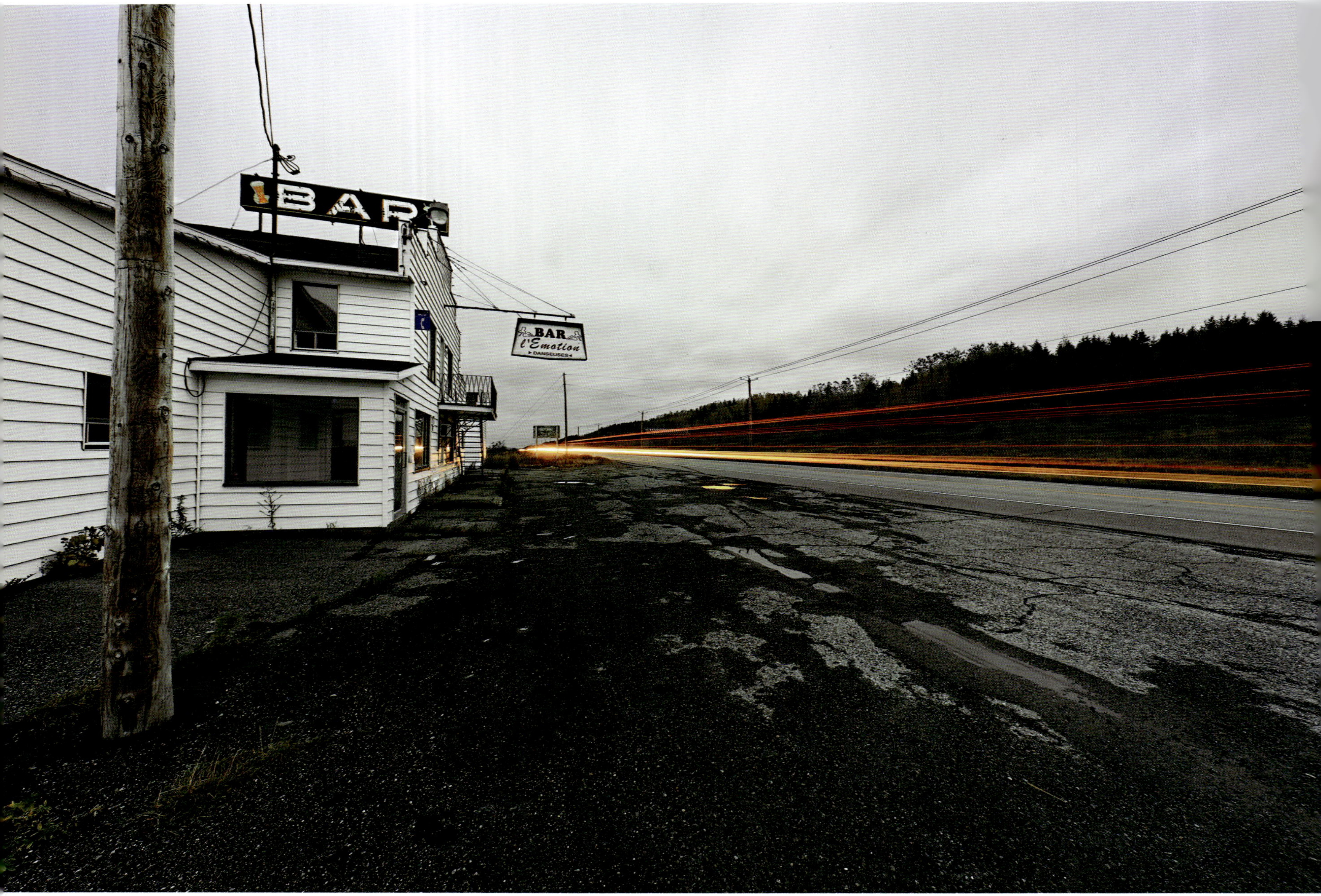

Strip Club, near Saint-Eloi-Station QC

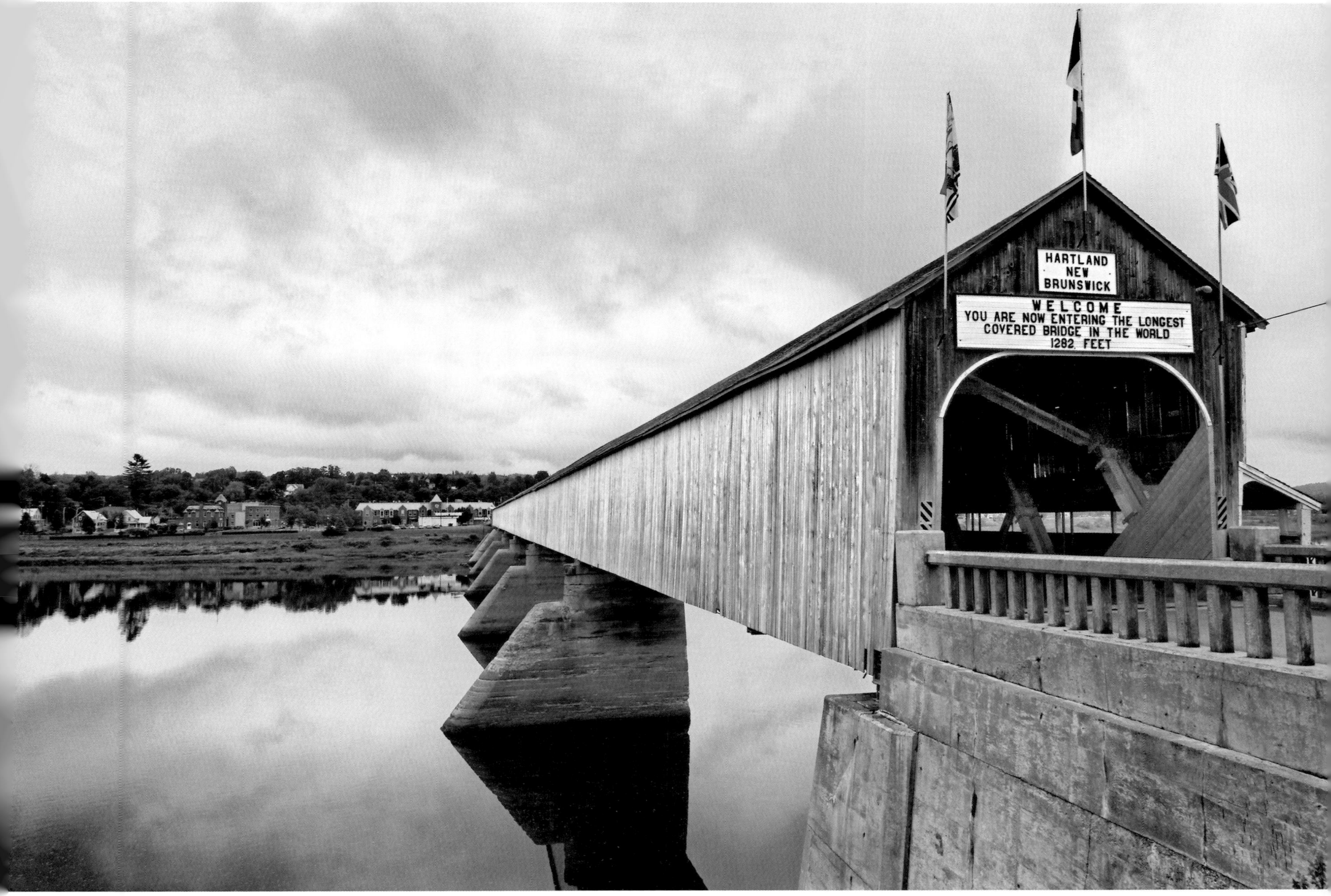

Longest Covered Bridge in the World, Hartland NB

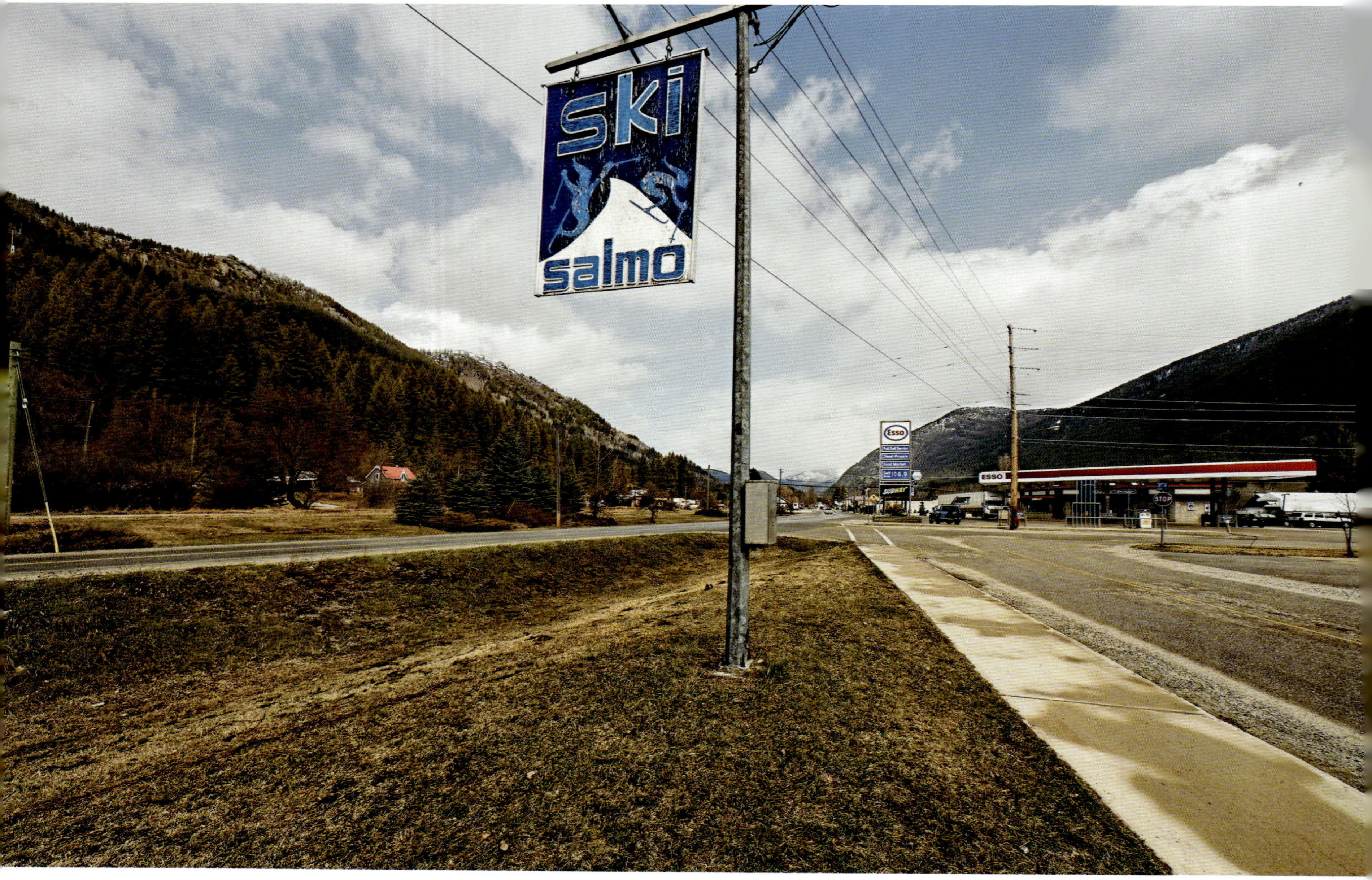

Ski Salmo, Salmo BC

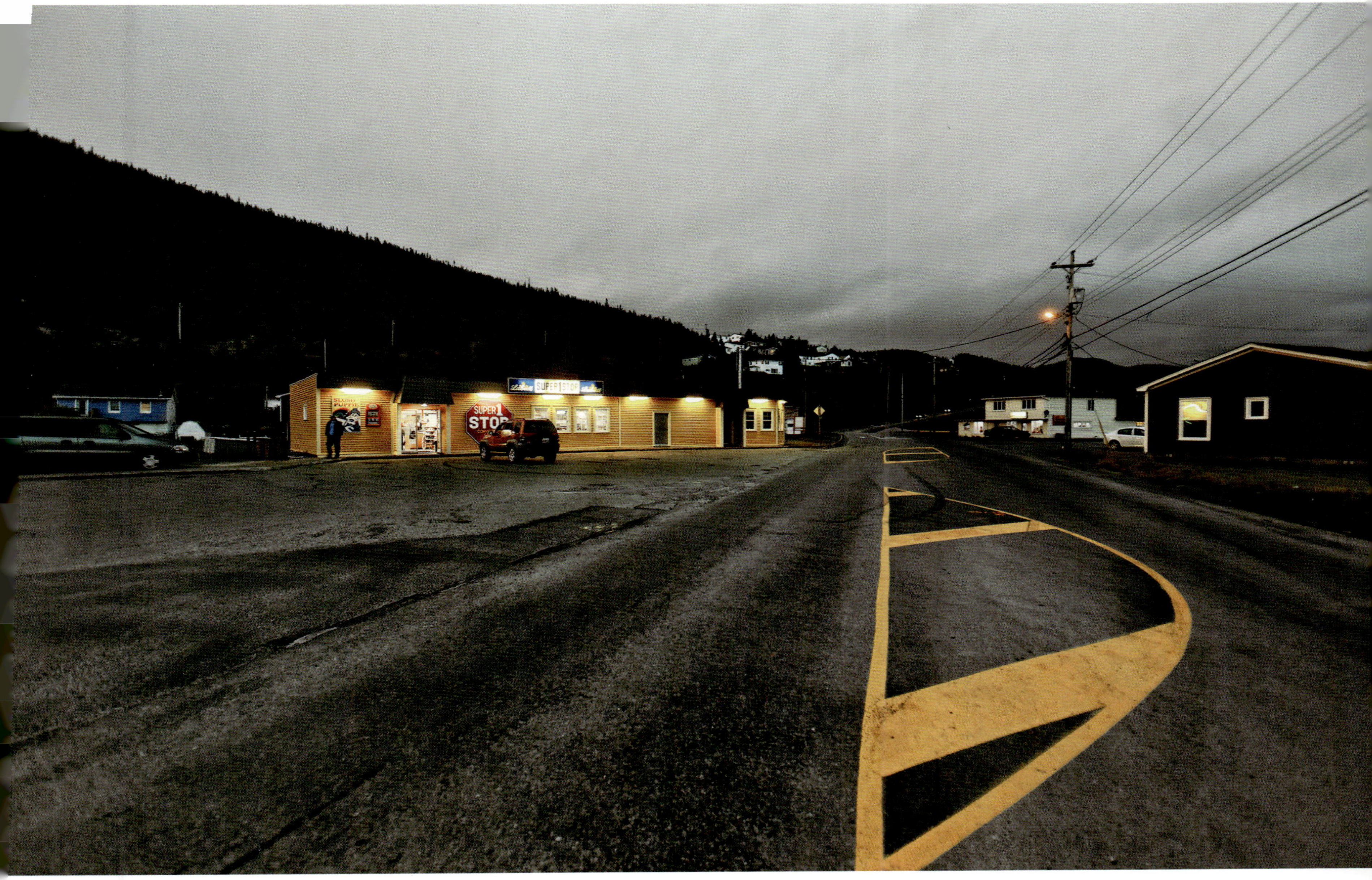

Convenience Store, Placentia NL

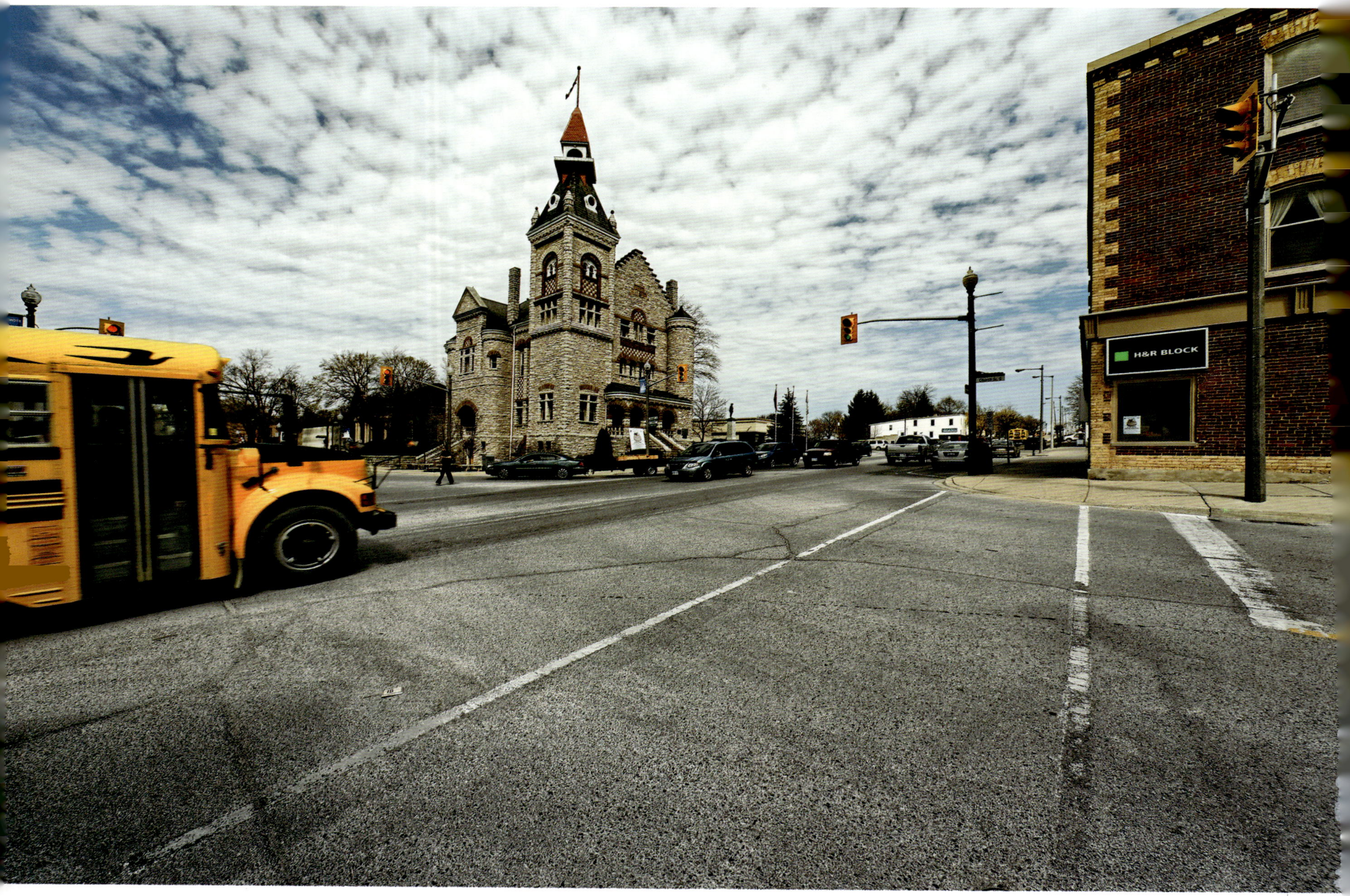

Town Hall, St. Marys ON

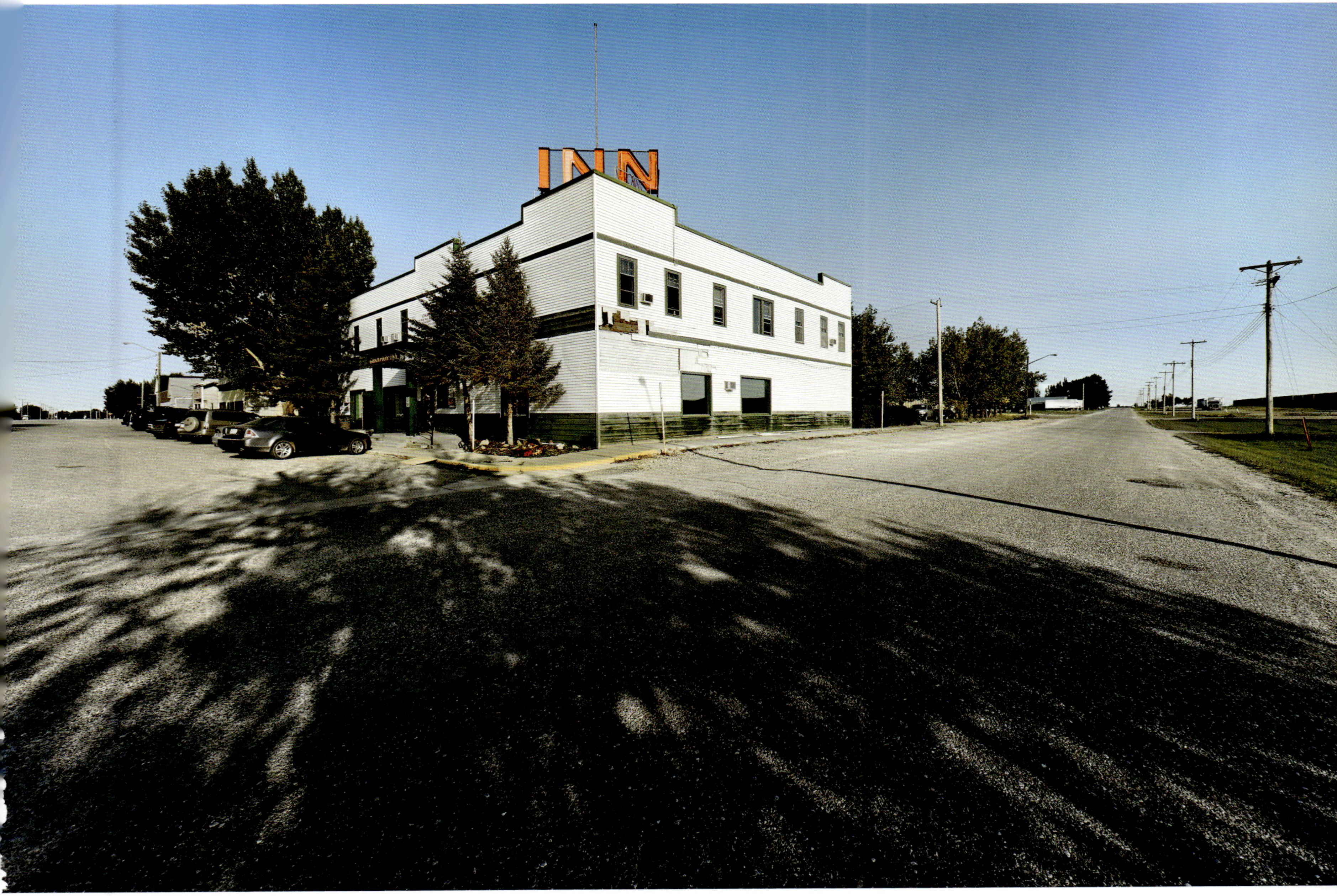

Champion Inn, Champion AB

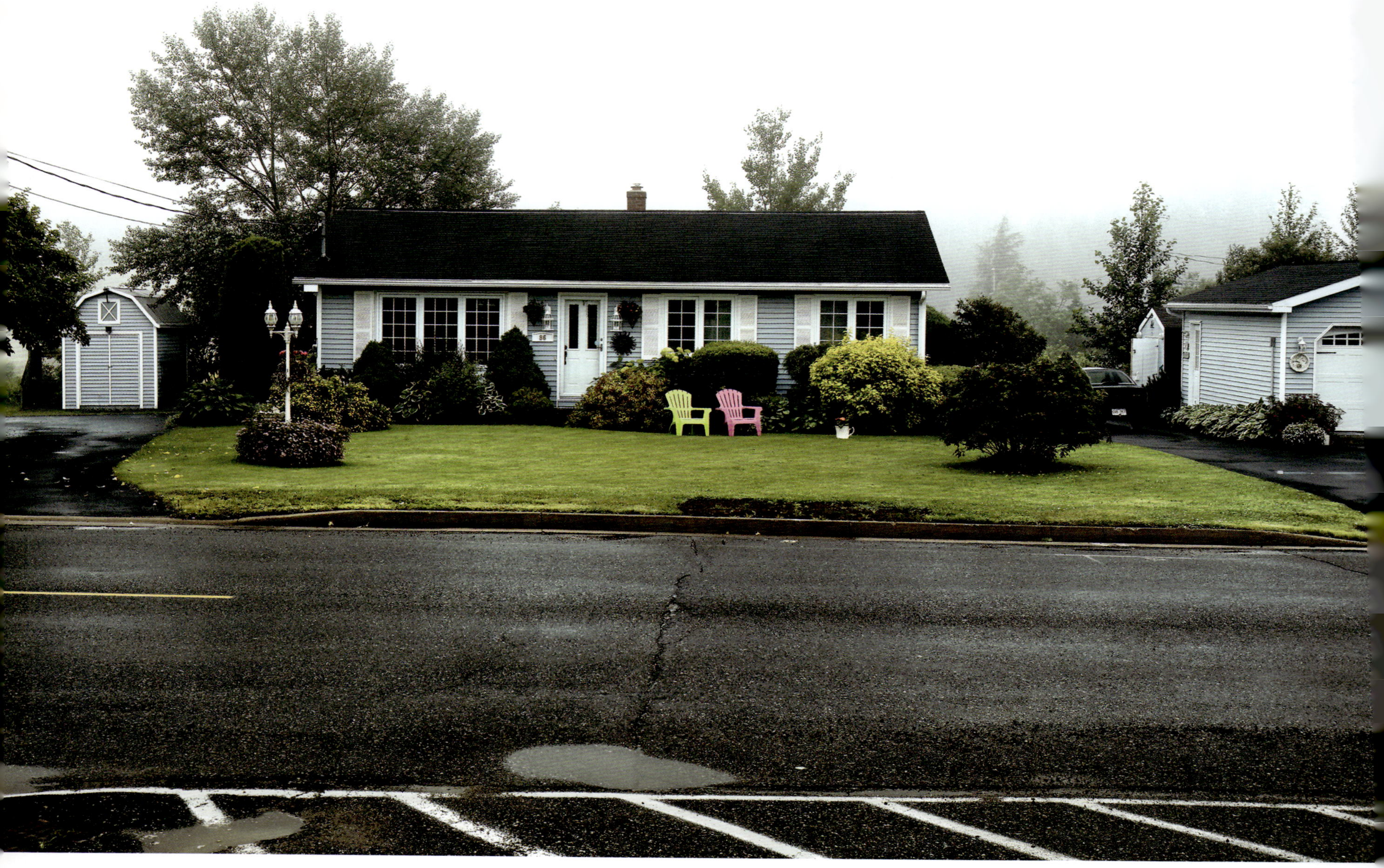

Two Chairs, St. George NB

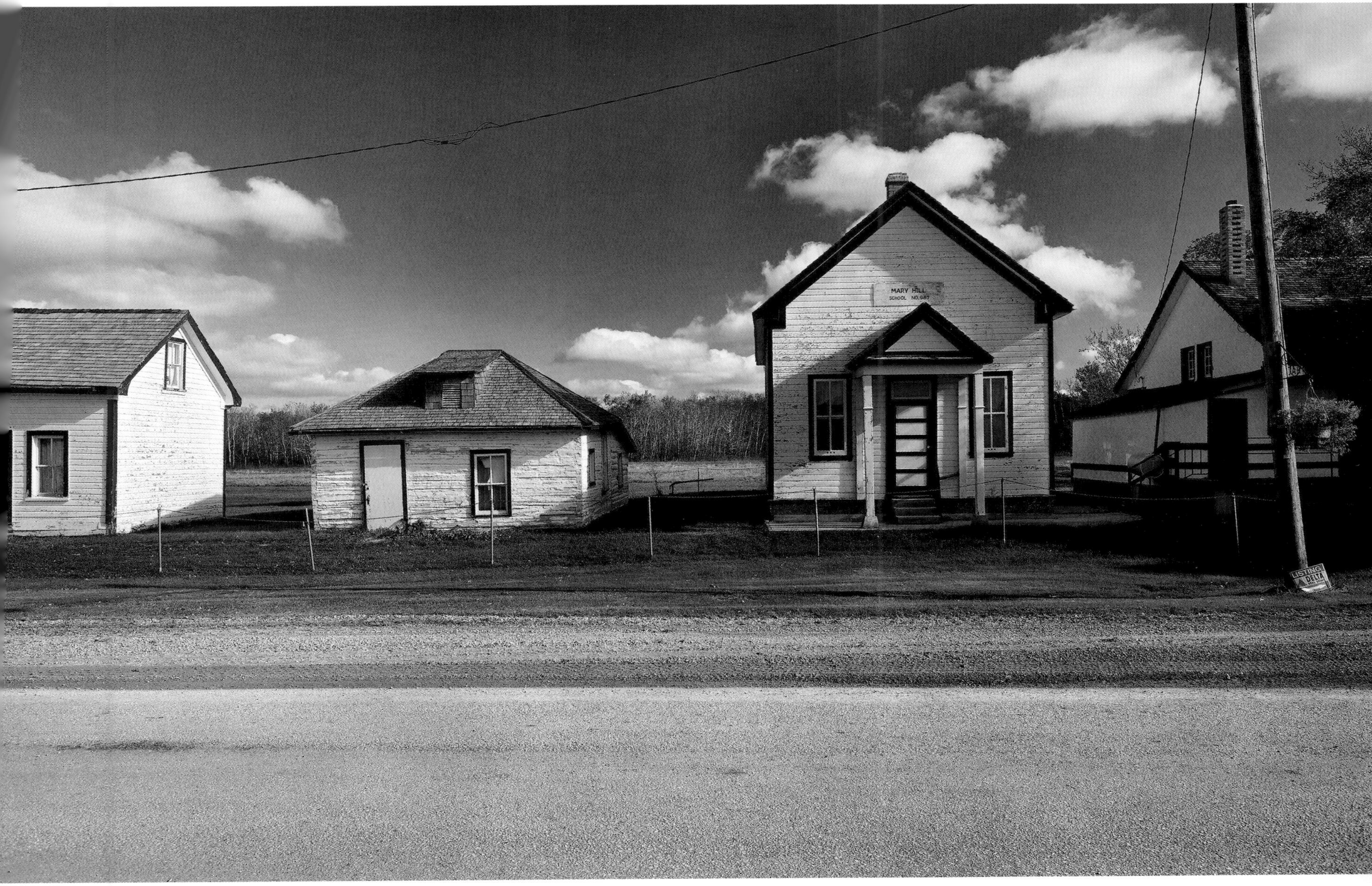

Lundar, Lundar MB

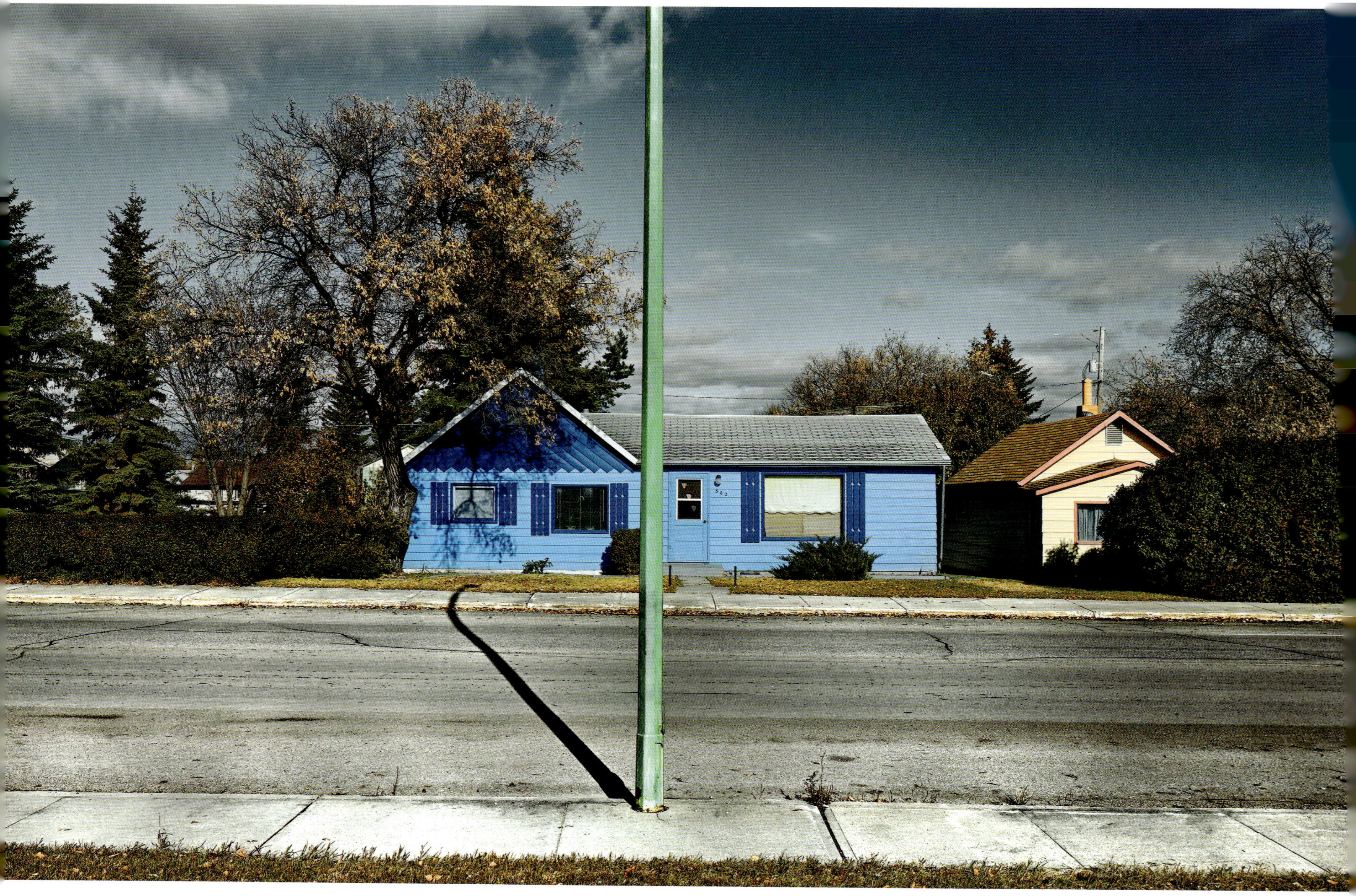

Blue House, Melville SK

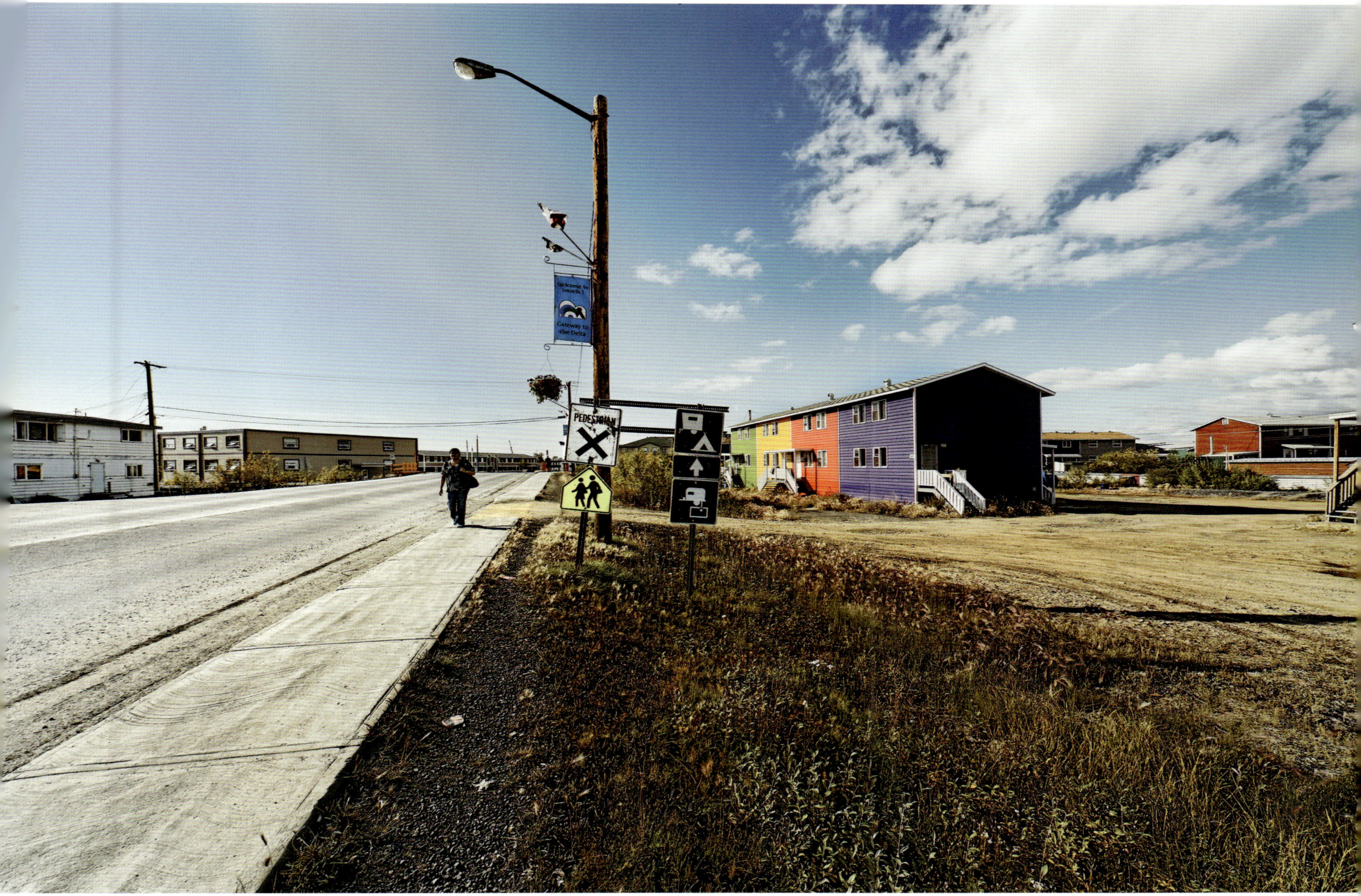

Mackenzie Road, Inuvik NT

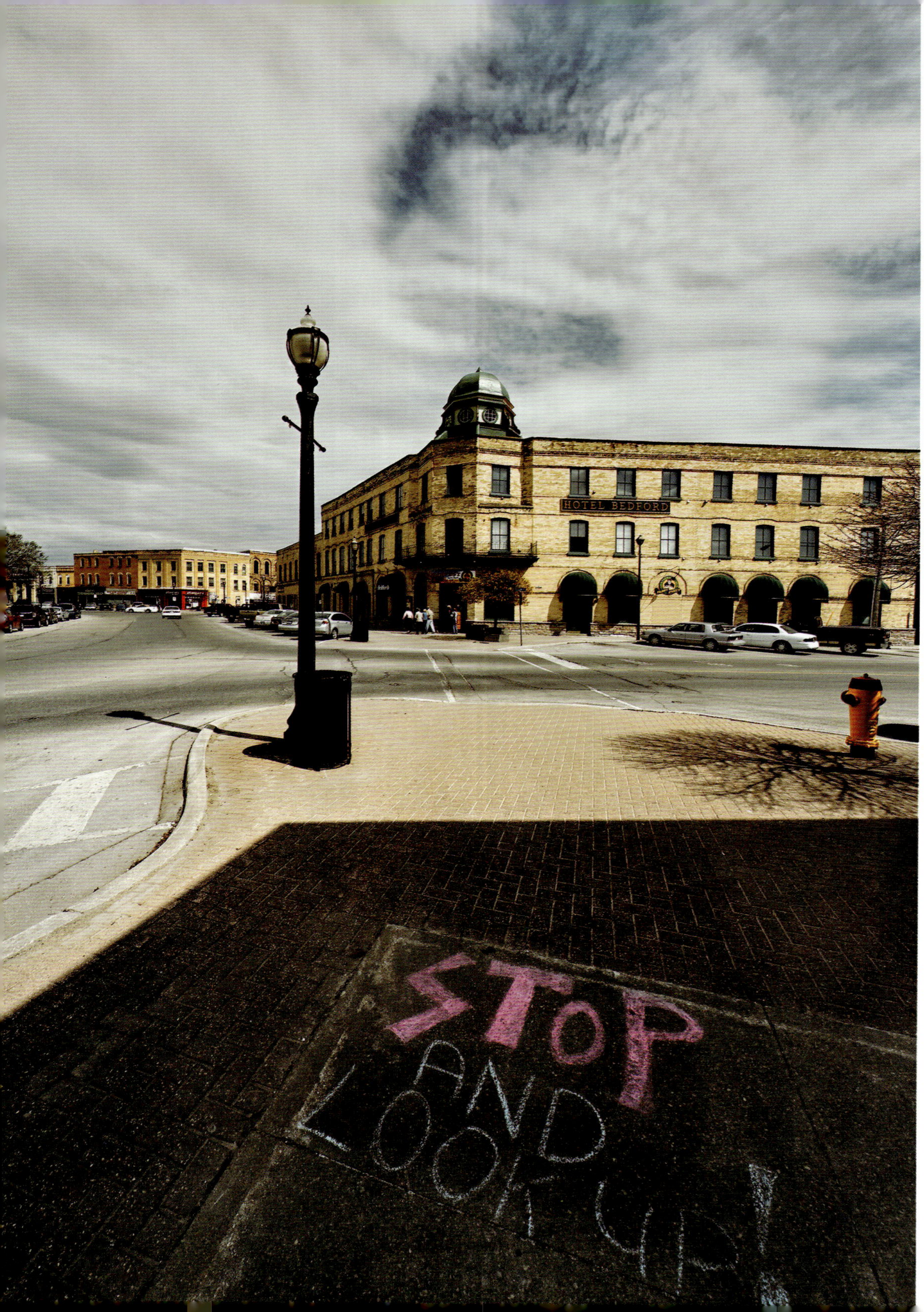

Stop and Look Up, Goderich ON

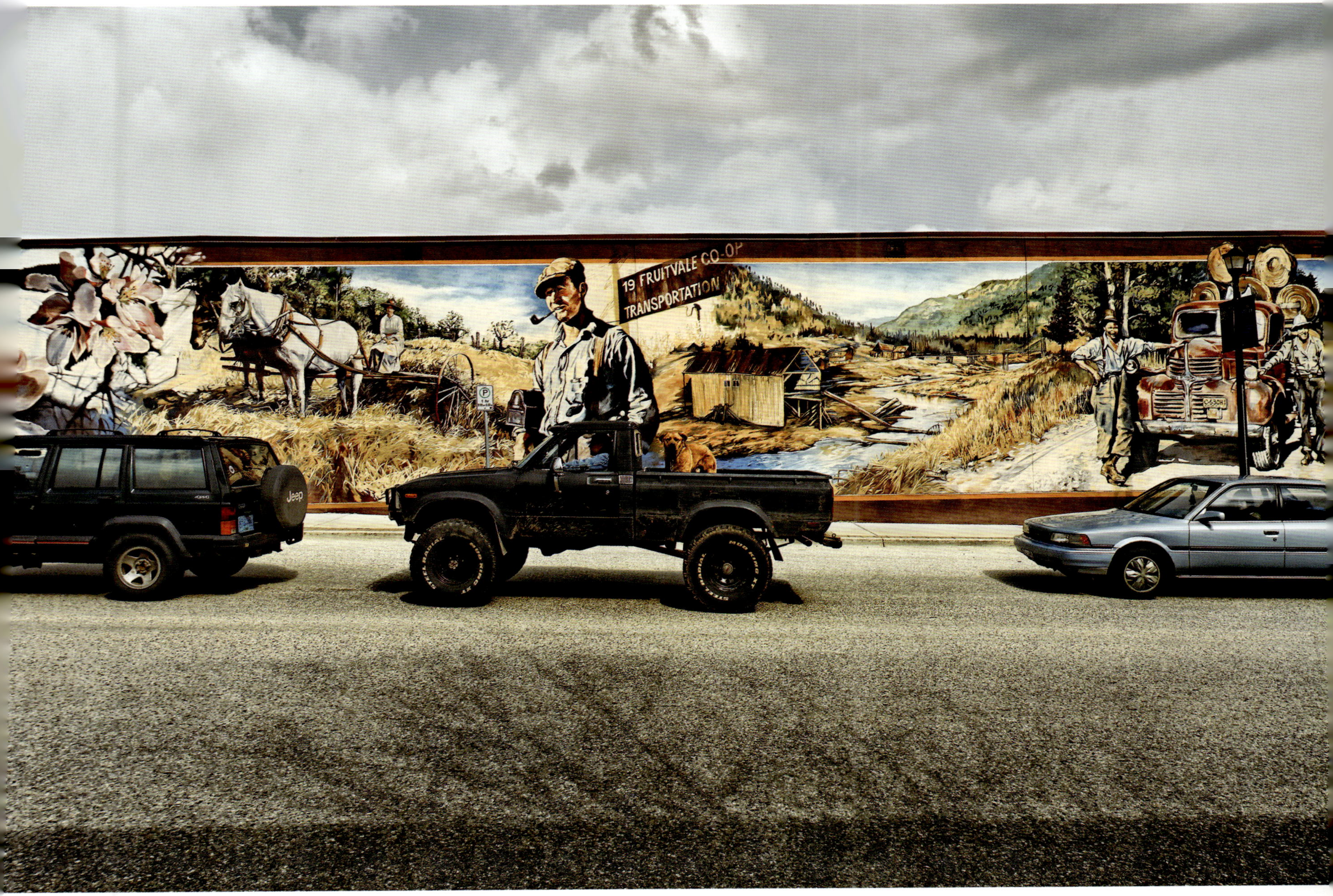

Mural, Fruitvale BC

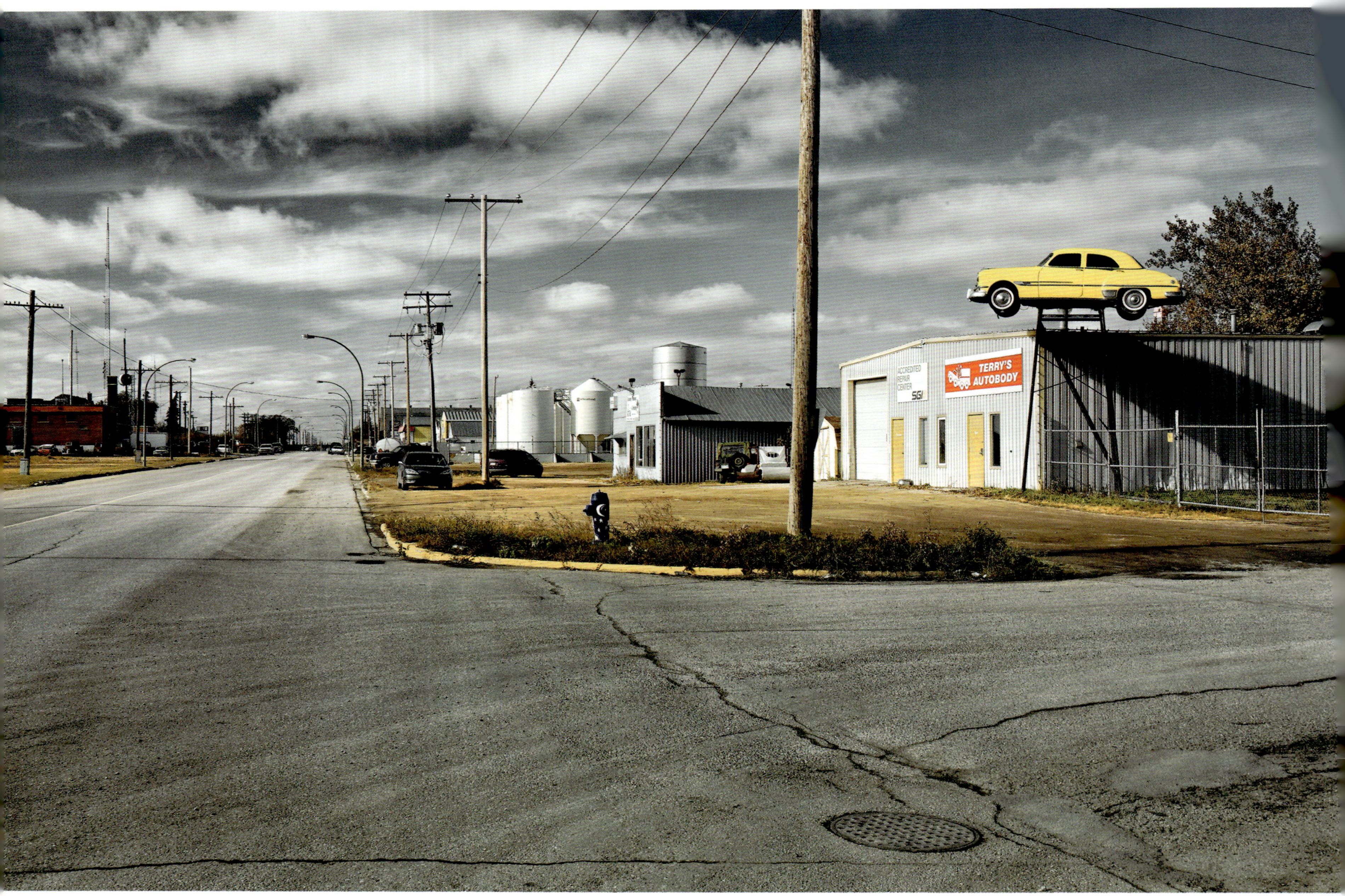

Terry's Auto Body, Melville SK

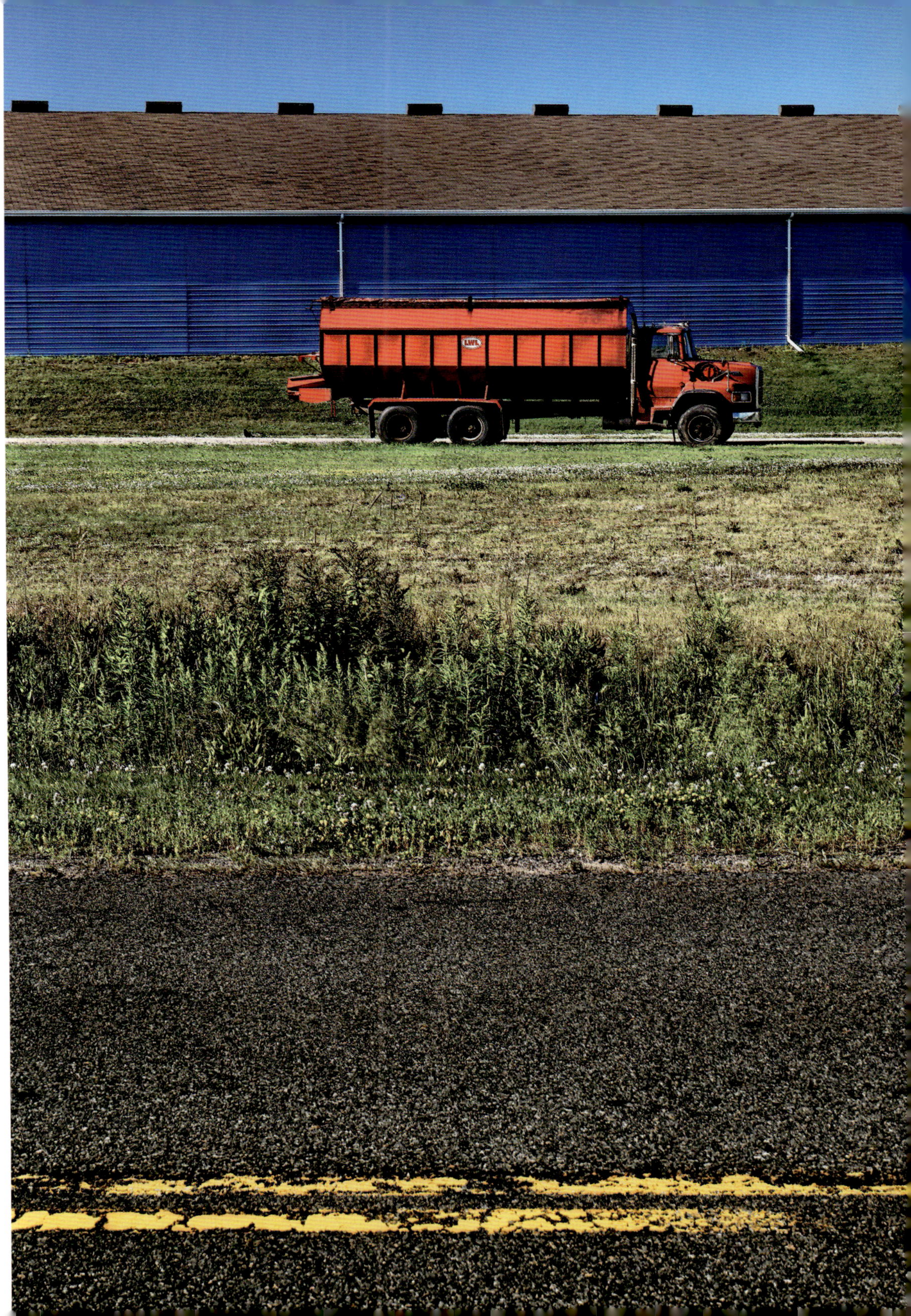

Red Truck, Canoe Cove PEI

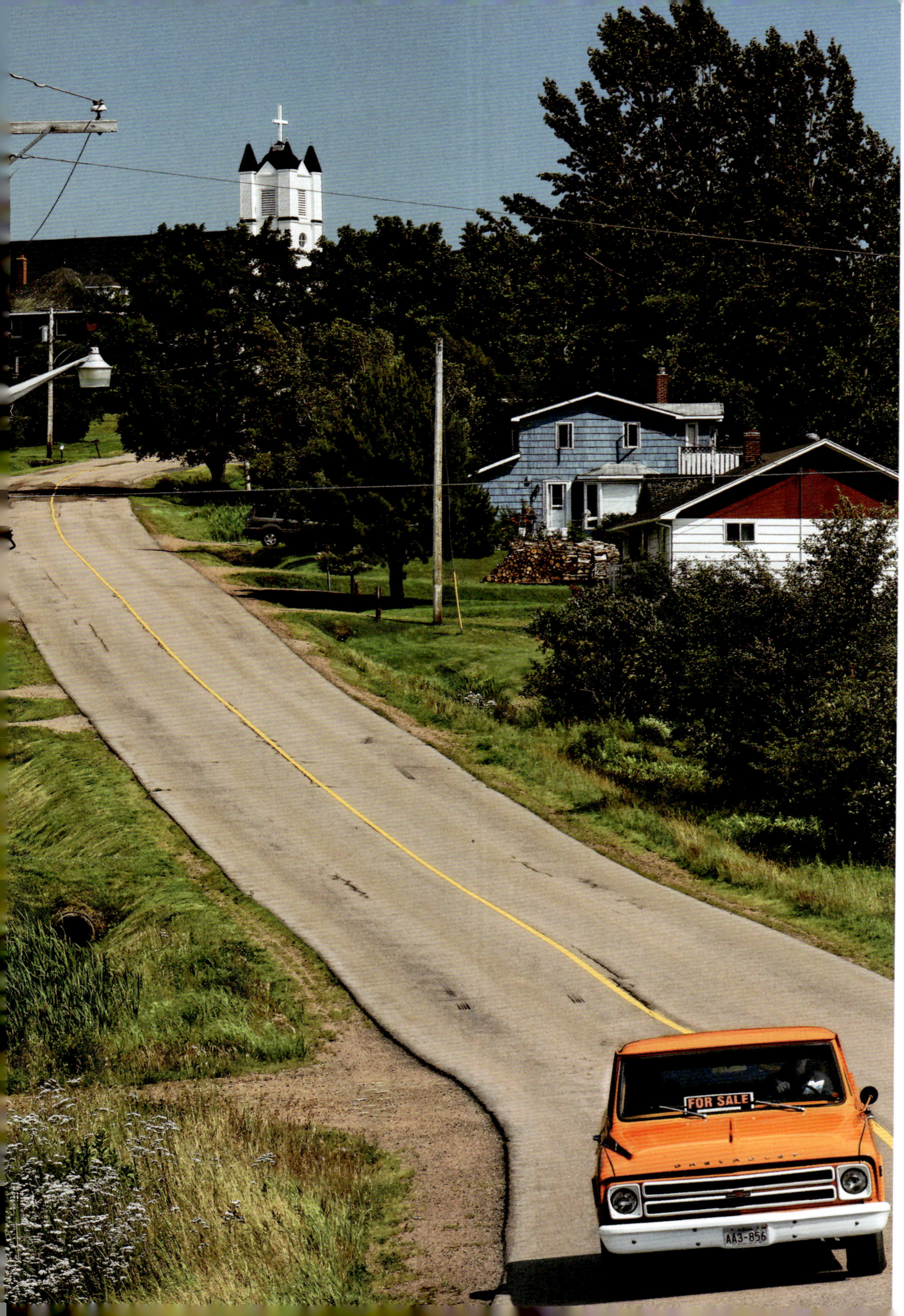

Orange Truck, Pré-d'en-Haut NB

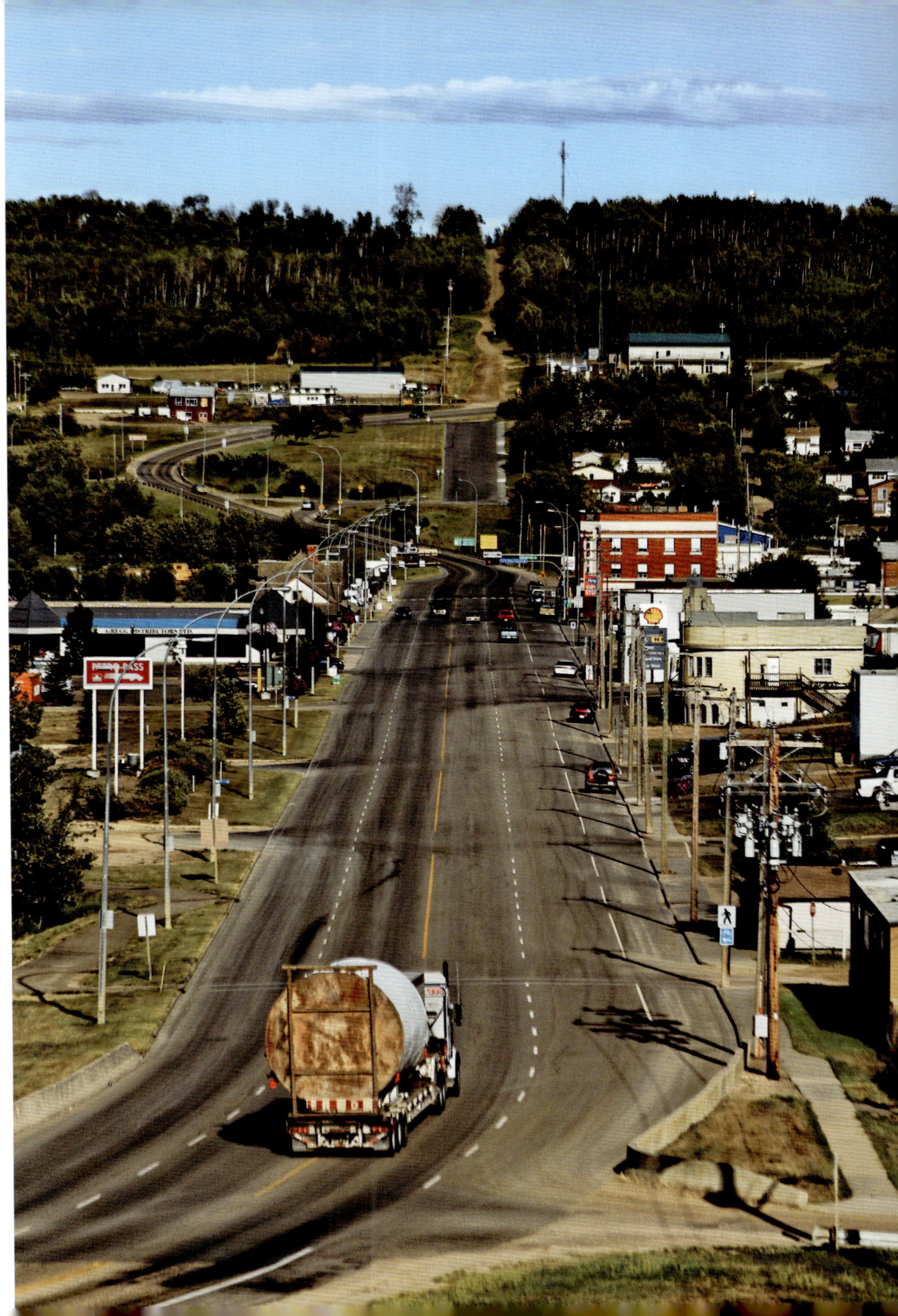

50th Avenue, Athabasca AB

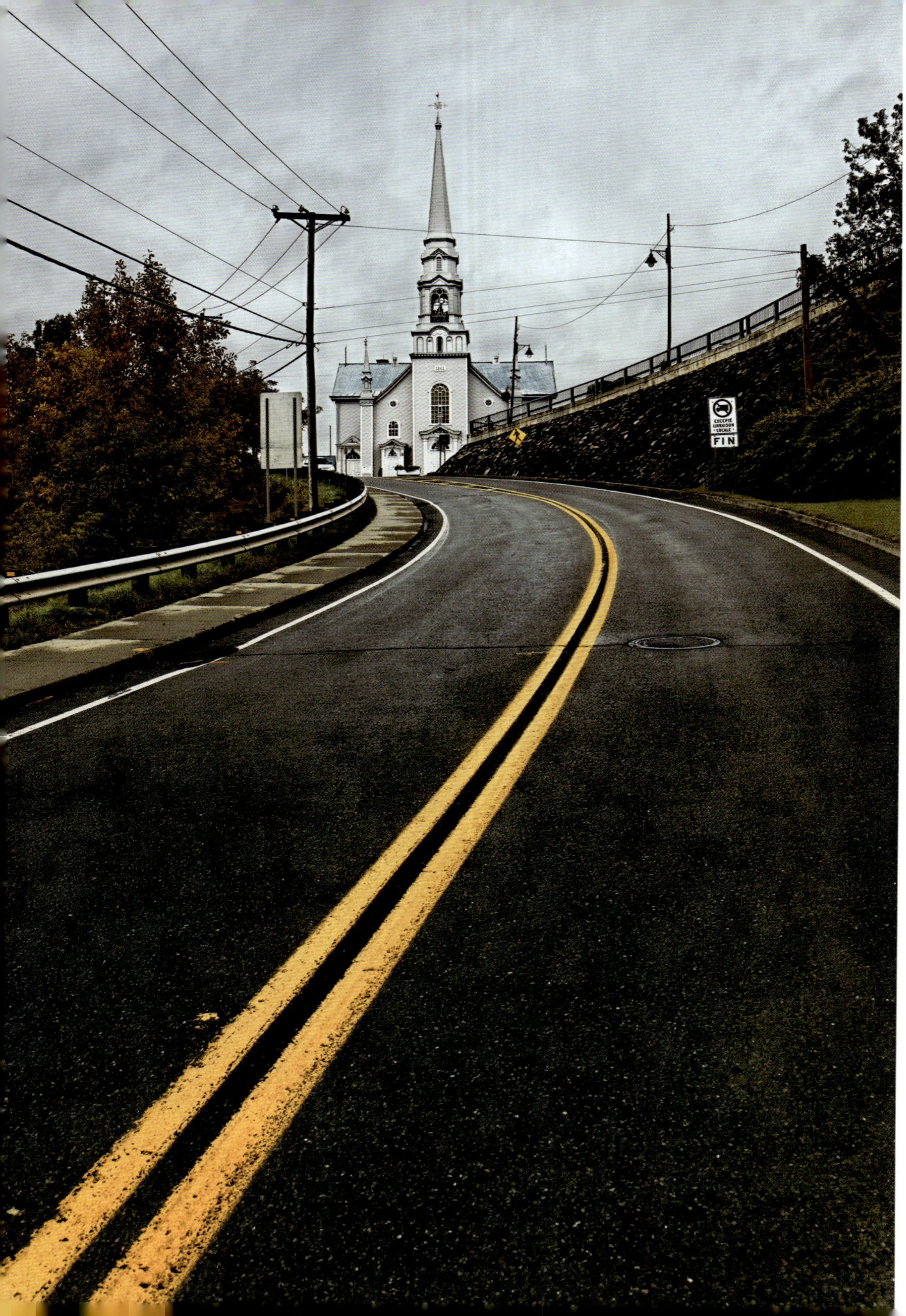

Saint Pacome, Saint Pacome QC

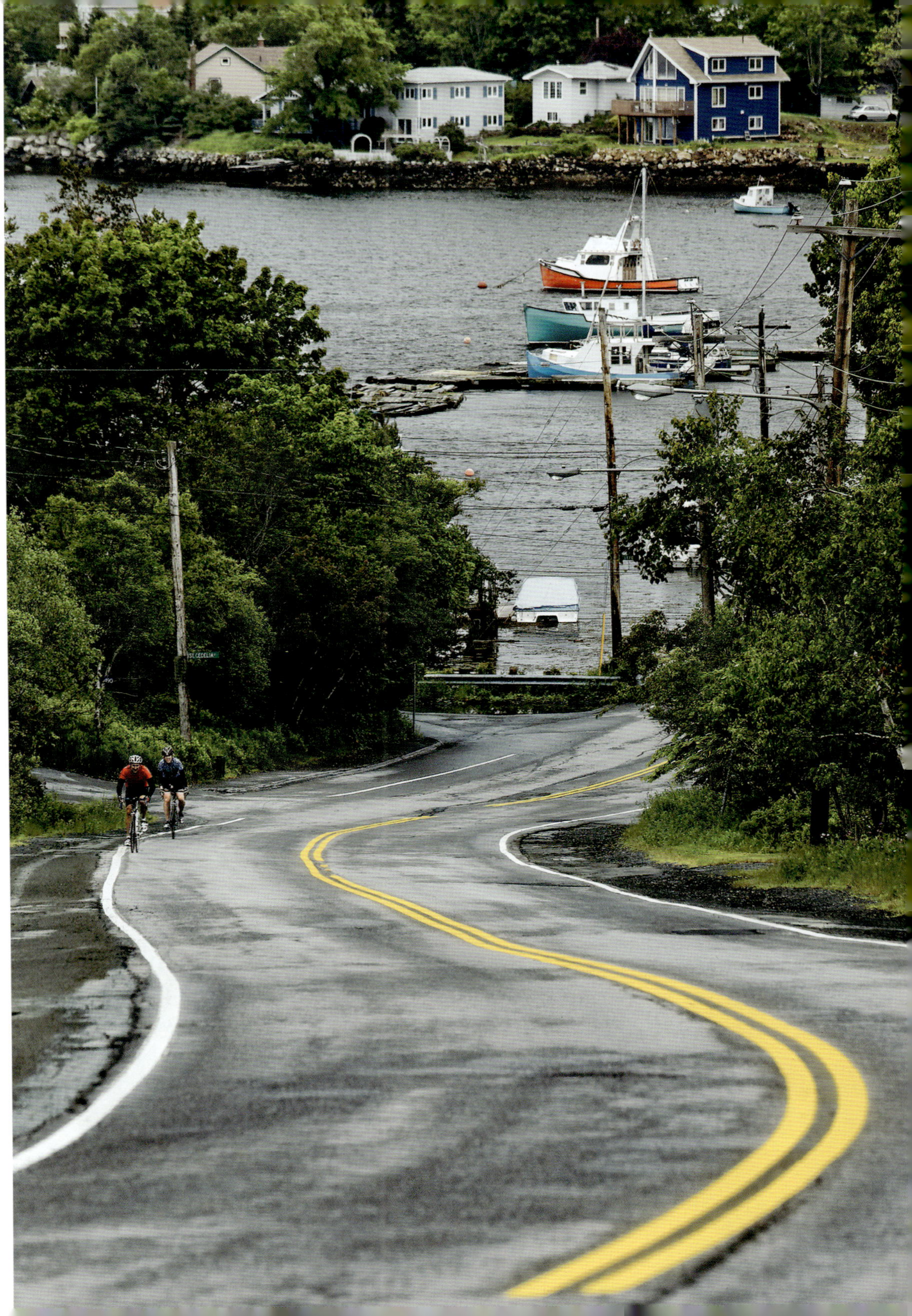

Purcell's Cove, Purcell's Cove NS

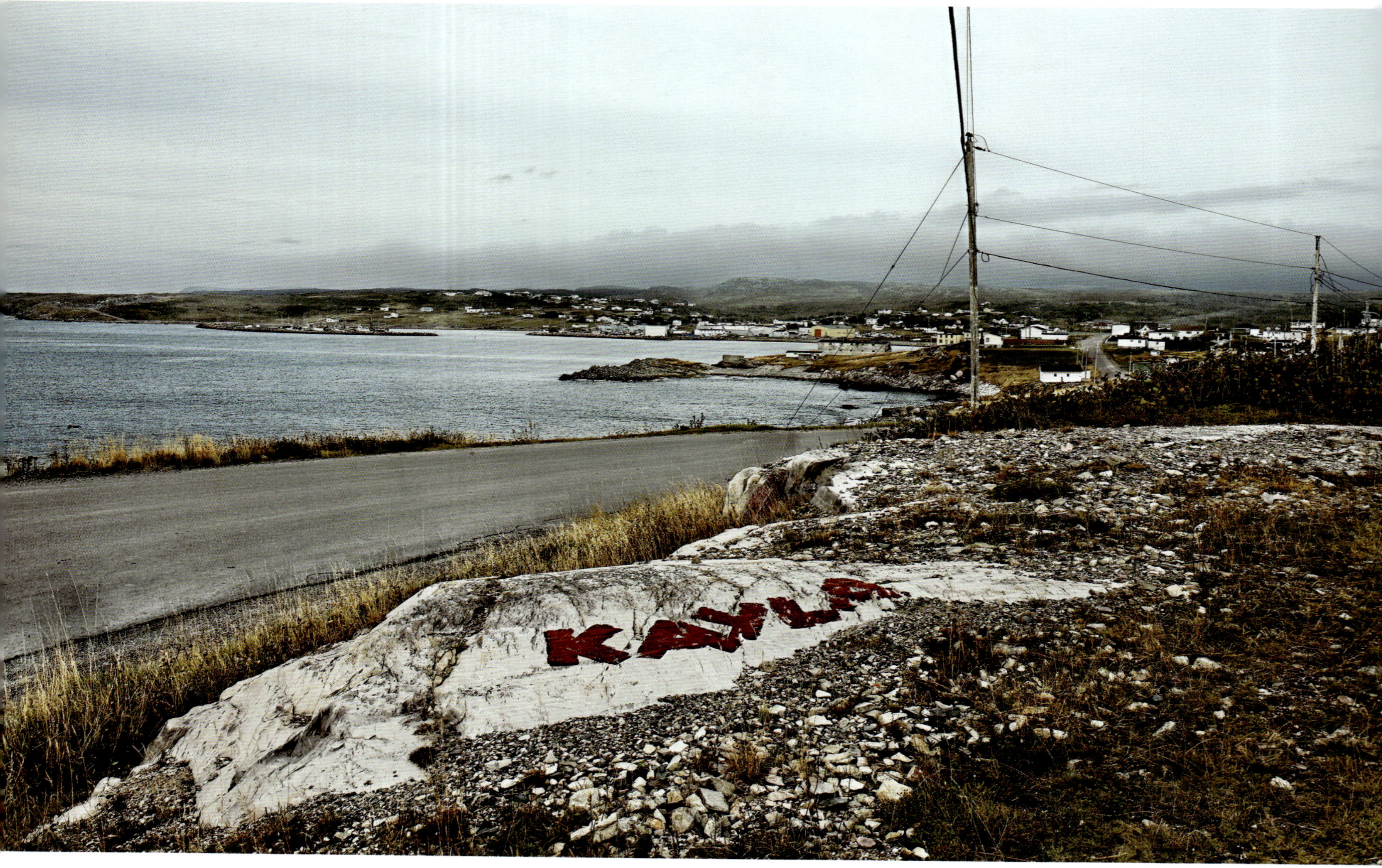

Kayla, Old Perlican NL

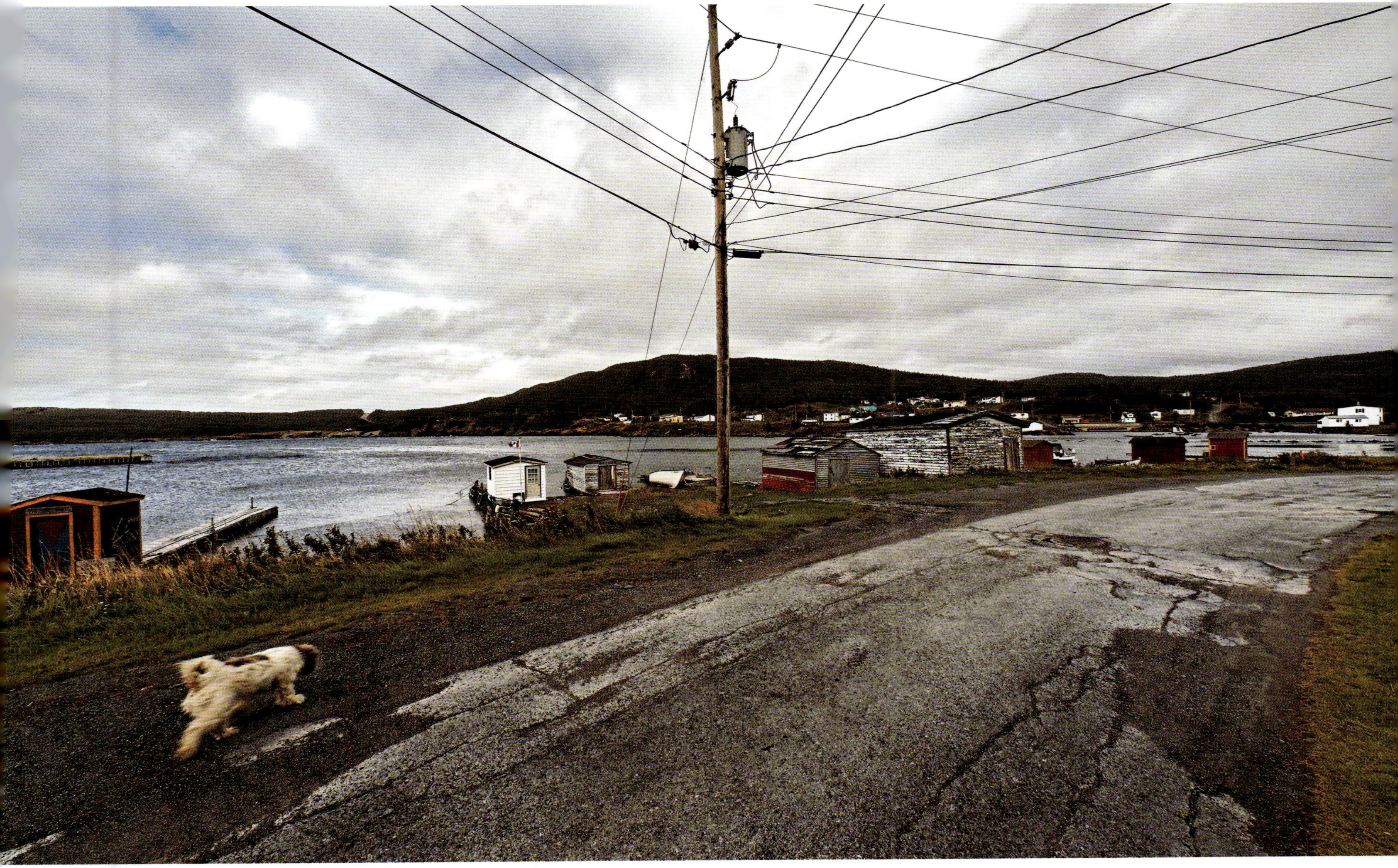

Dog, New Perlican NL

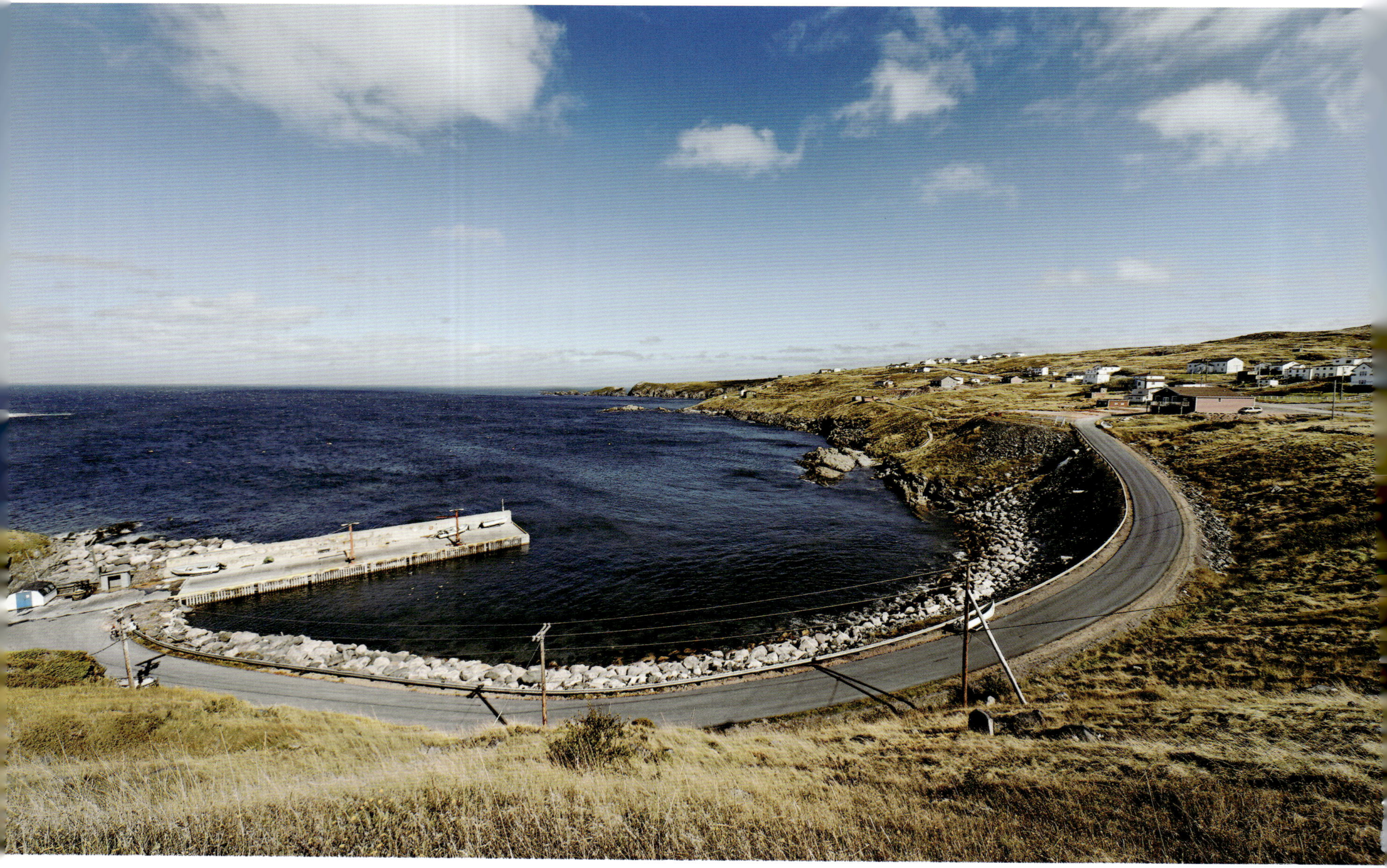

Grates Cove, Grates Cove NL

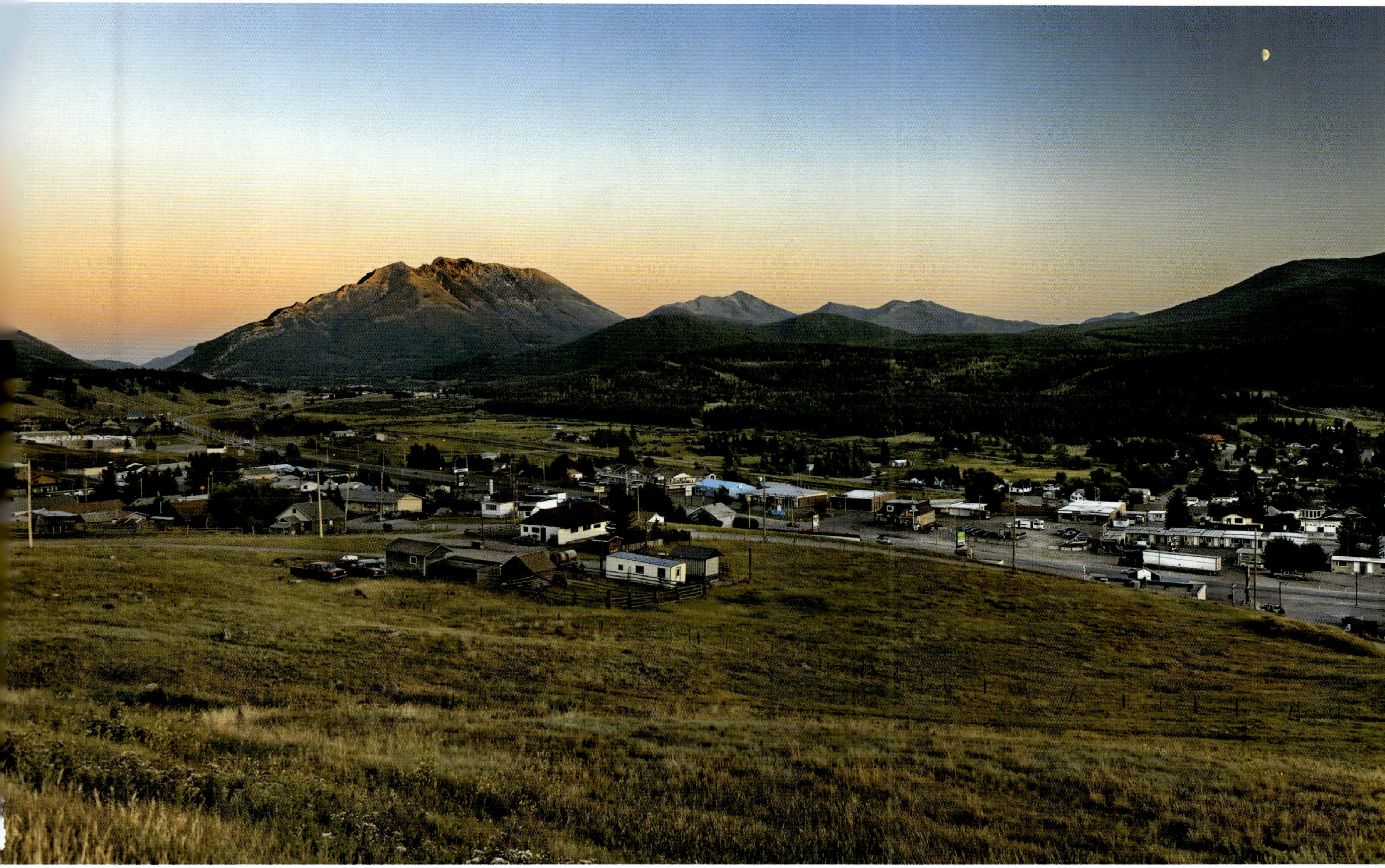

Coleman, Coleman AB

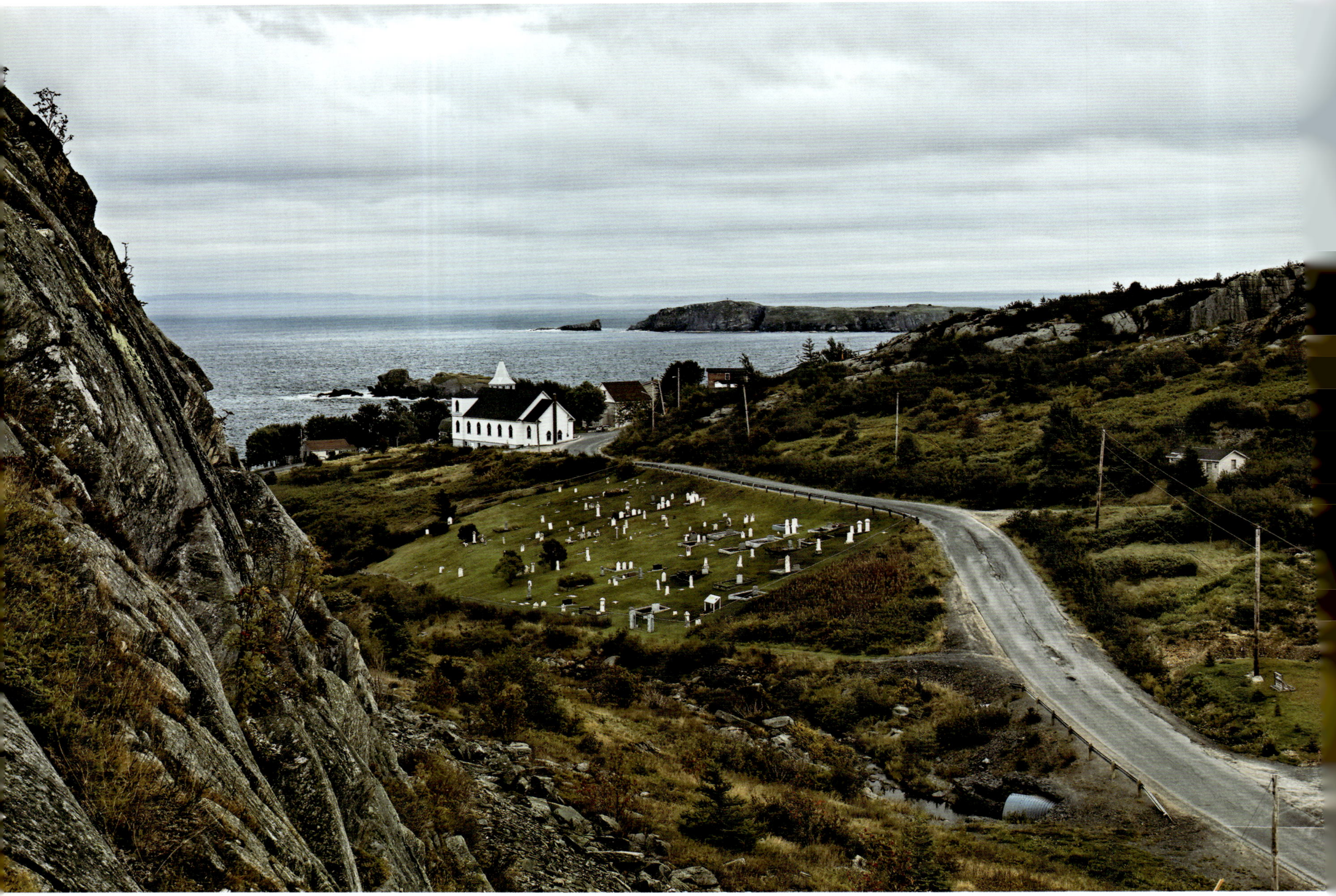

Freshwater, Freshwater NL

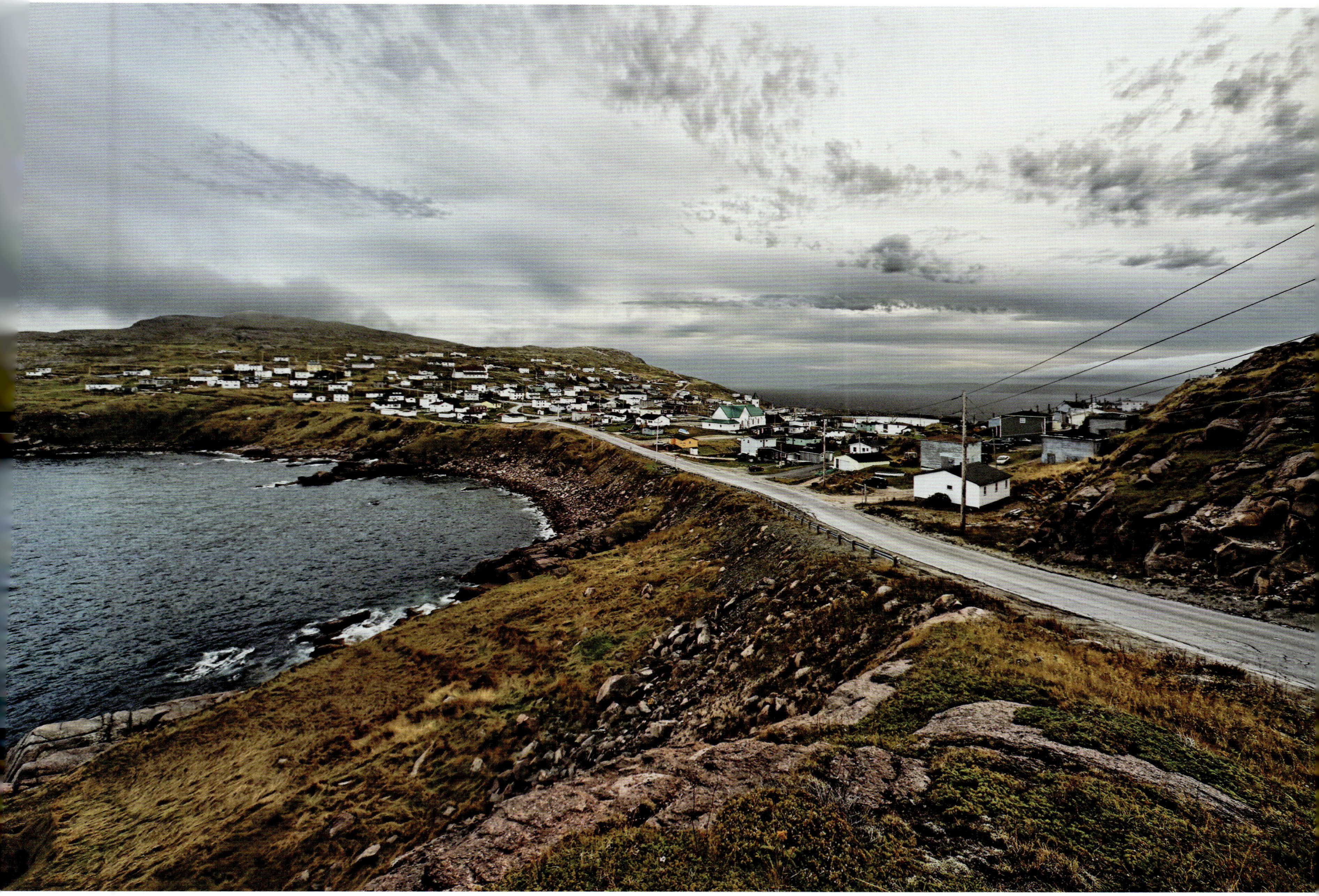

Bay de Verde, Bay de Verde NL

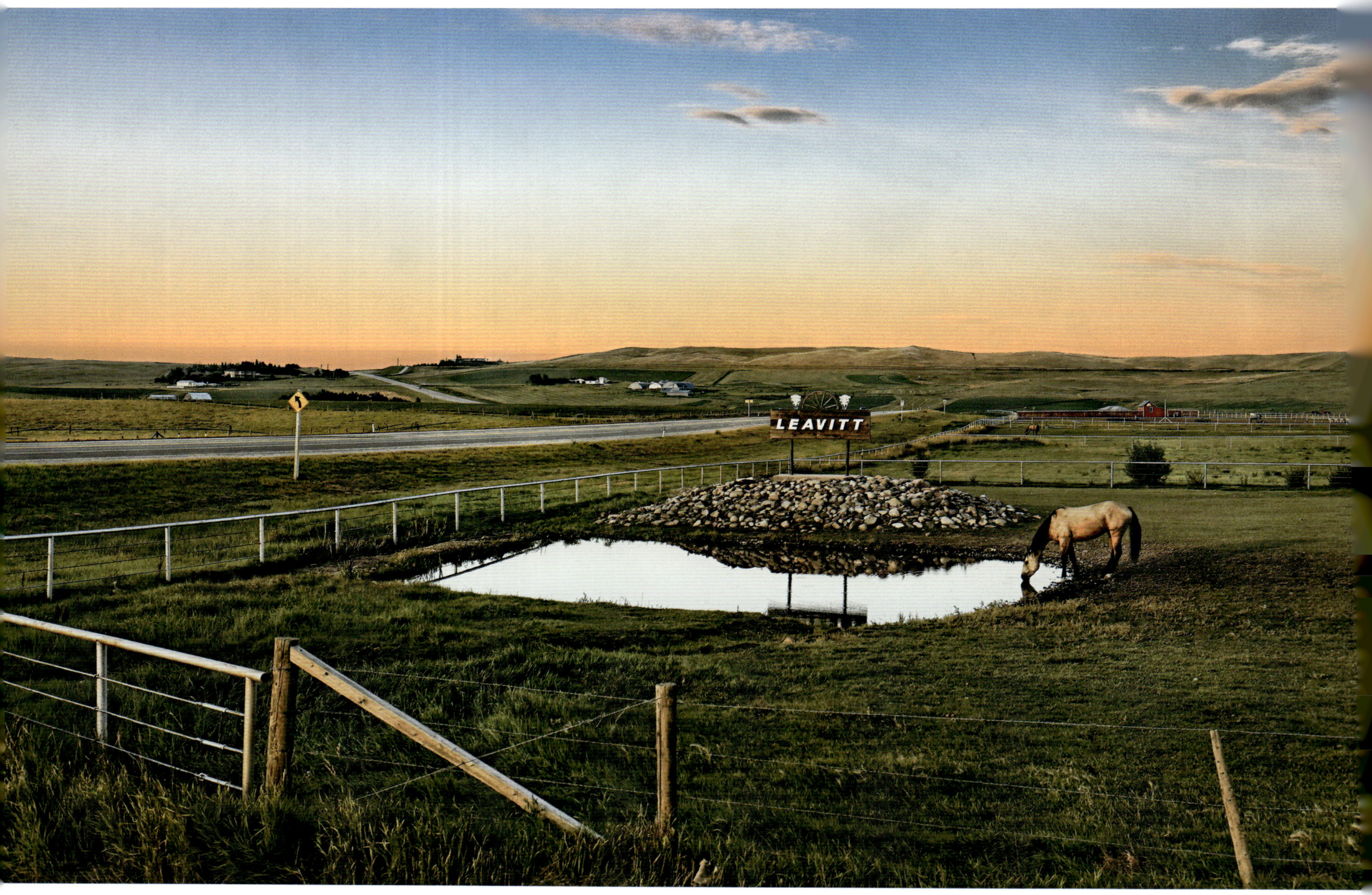

Horse Drinking, Leavitt AB

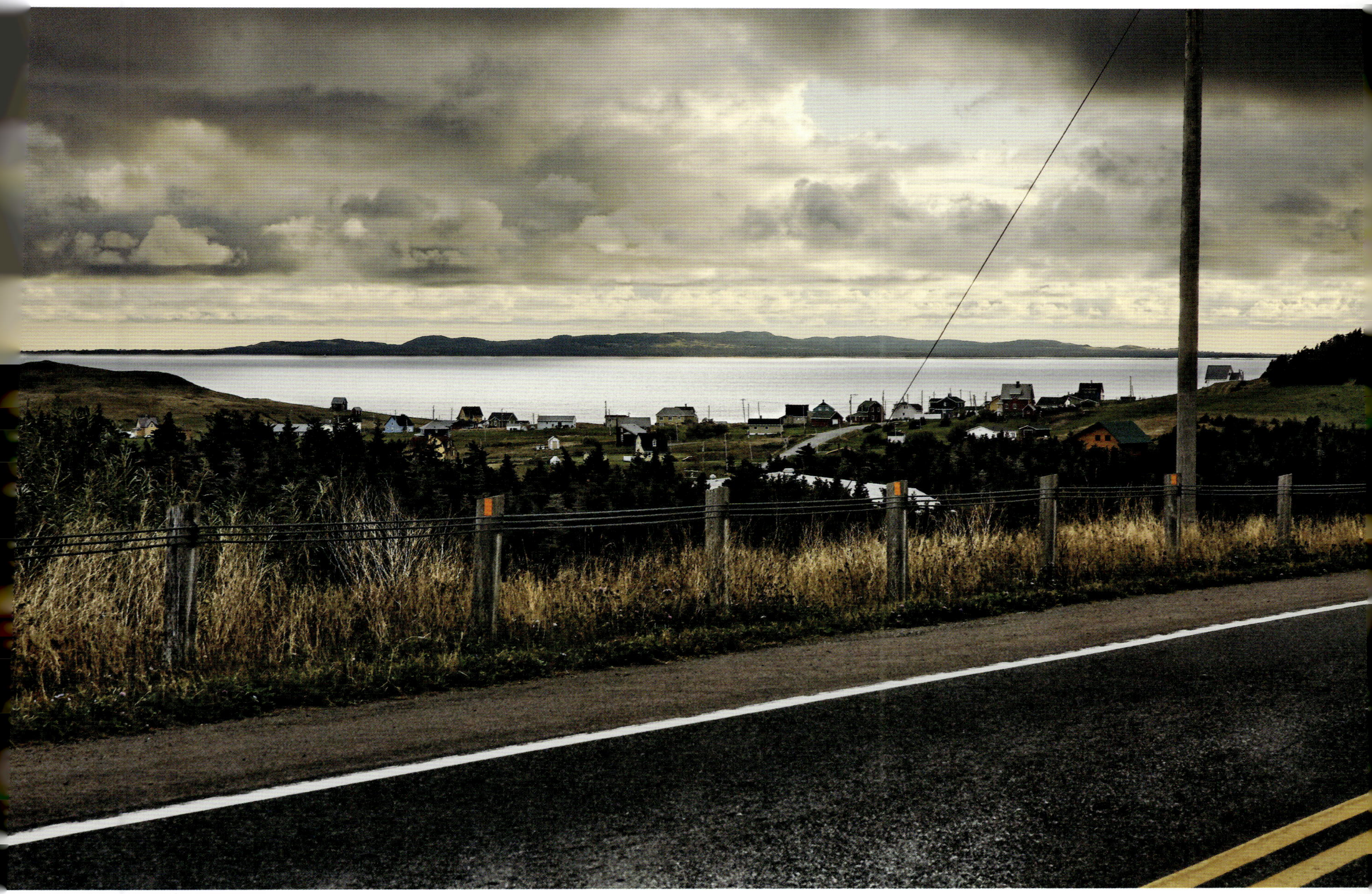

Havre-aux-Maisons, Havre-aux-Maisons (Magdalen Islands) QC

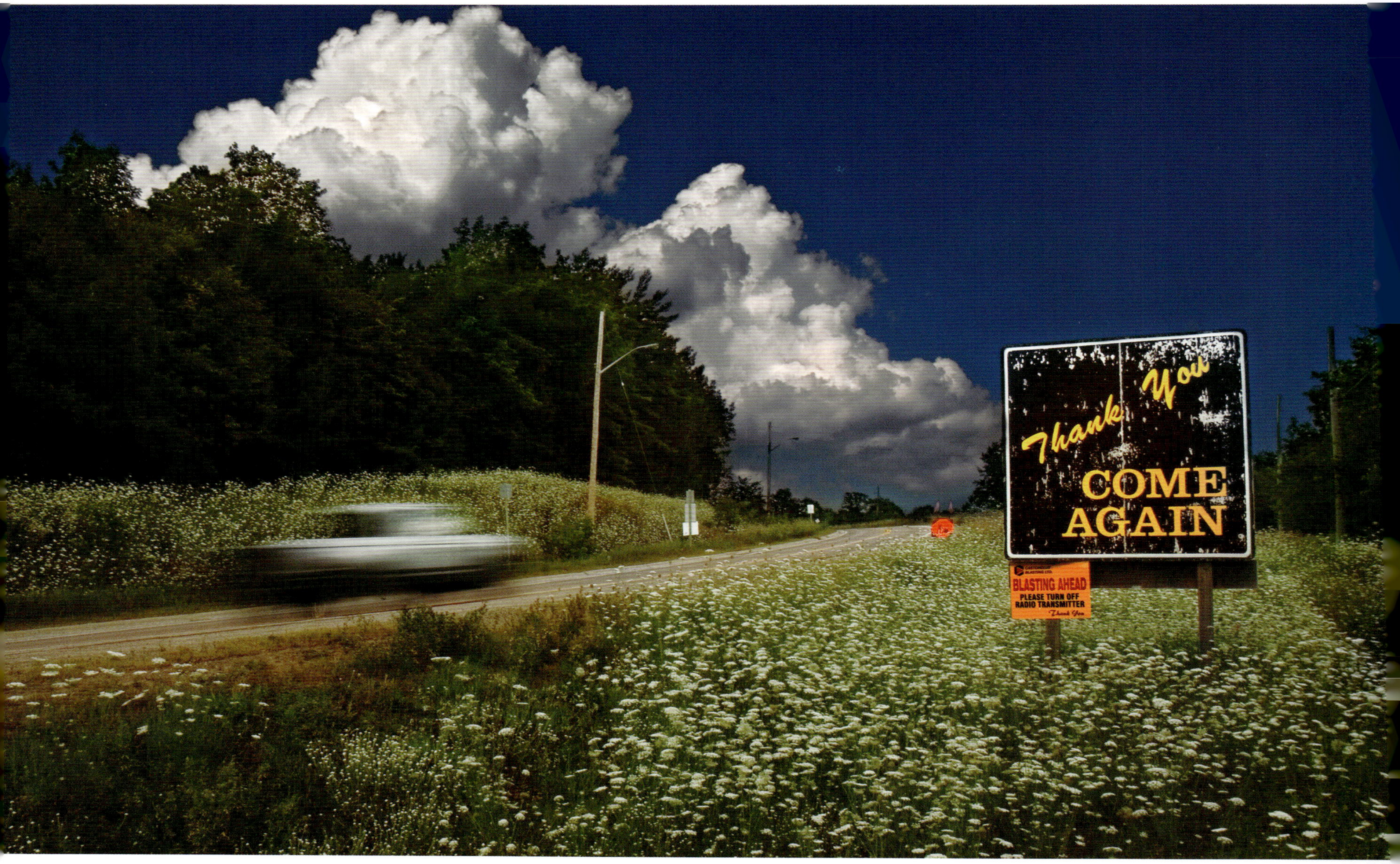

Leaving Eganville, Eganville ON

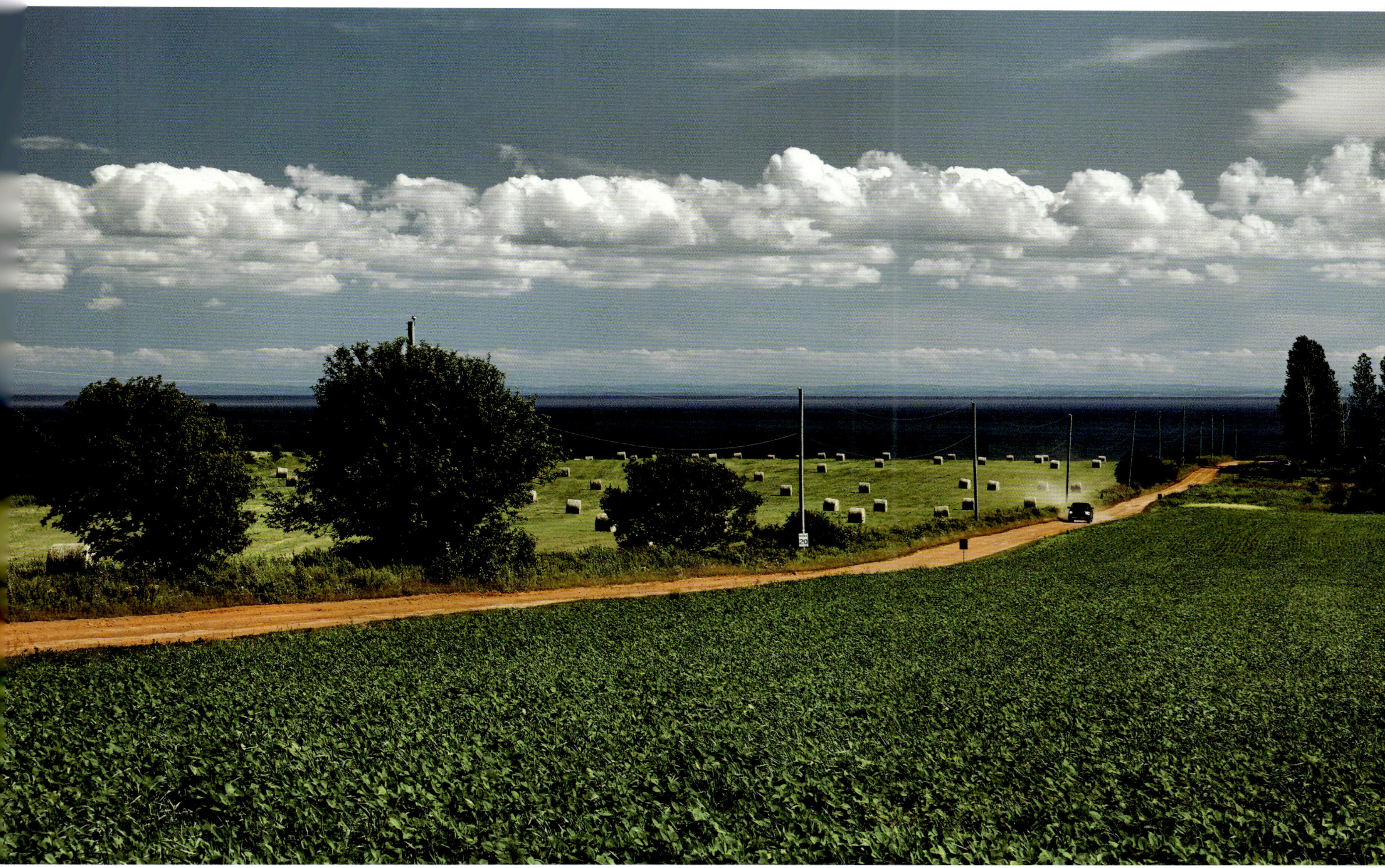

Argyle Shore, Argyle Shore PEI

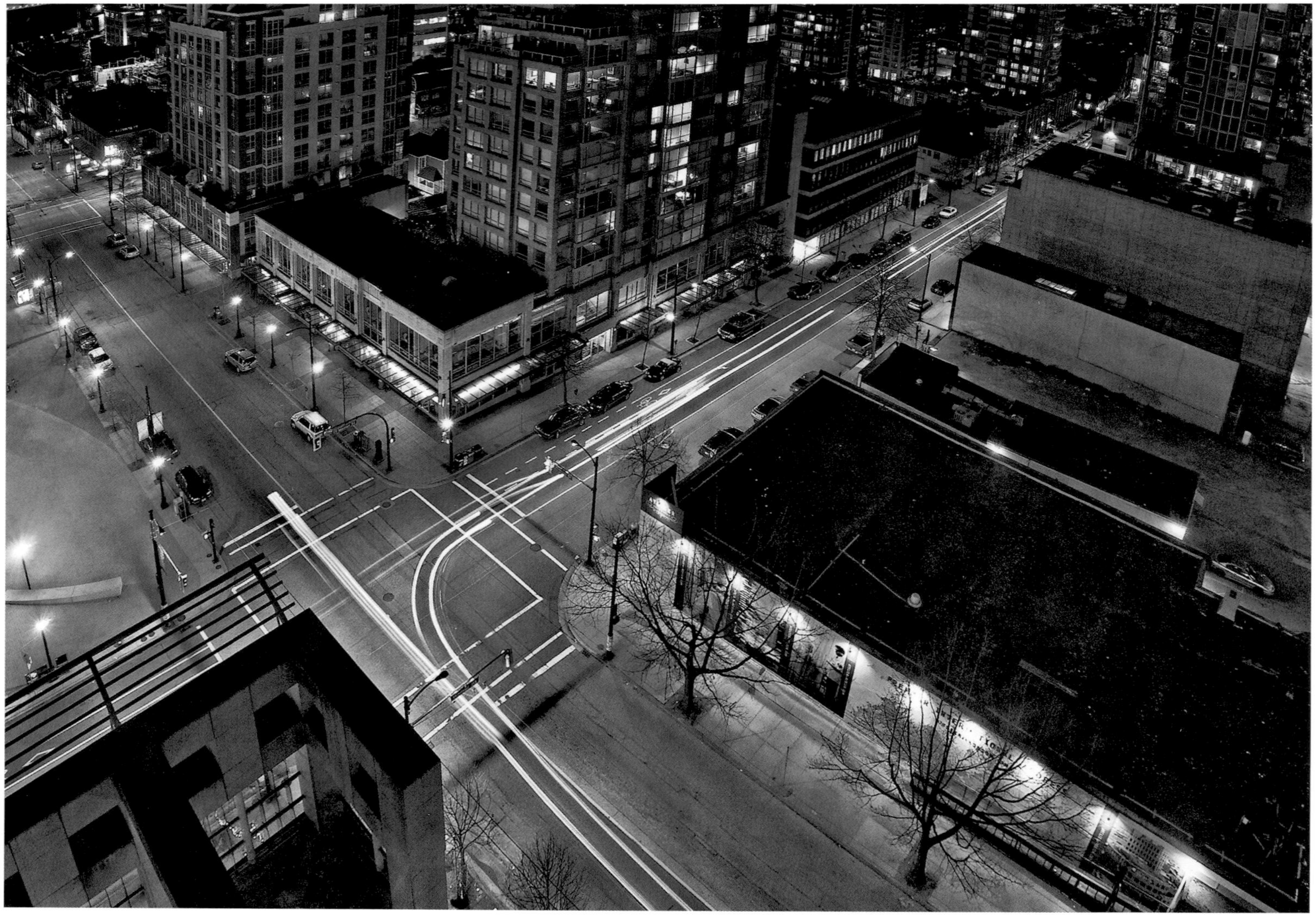

City Lights, Vancouver BC

Cities

IN EMPTY SPACES AND RURAL AREAS the road is a sign of intrusion: a provocation, a declaration, like graffiti on a wall, that "man is here." Where the population is sparse, roads stand in contrast to the landscape, creating dramatic visual tension.

It's different in cities where roads are the circulatory system and backbone, the warp and weft, of the urban landscape. The dramatic tension is gone. And so a superhighway can look serene in an urban setting (*On Ramp*, p. 161); the flow of urban traffic (*City Lights*, p. 152; *Midnight Bus*, p. 168) seems as inevitable (and, in an unexpected way, as beautiful) as a river flowing to the sea; the laws of nature seem mimicked by man-made laws expressed in traffic lights and speed limit signs (*McKnight Boulevard*, p. 180).

Elevated roadways—bridges and viaducts—are the high mountain passes of the urban landscape. Like mountain passes, bridges have strategic significance: they are focal points of traffic bulletins on radio and television (as commuters in Ottawa, Montreal, or Vancouver will attest). And many of them rise above their mundane function as transportation links to become definitive urban landmarks. The Lions Gate Bridge (*Lions Gate*, p. 160) spanning Burrard Inlet—once a private toll bridge owned by the Guinness family of Irish beer-making fame—is synonymous with Vancouver. The Bank Street Bridge (*Bank Street Bridge*, p. 164) over the Rideau Canal is one of the most photographed structures in Ottawa. Some city bridges have whimsical nicknames—I think particularly of the span over the West Arm of Kootenay Lake in Nelson, British Columbia, built in 1957 to replace a cable ferry. For reasons apparent from the photograph (*Motel*, p. 157) the bridge is known locally as "Bob," short for "big orange bridge." And some bridges have become the stuff of legend. It is said of the University Bridge over the South Saskatchewan River in the heart of Saskatoon (*University Bridge*, p. 165) that during its construction in the early 1900s a worker fell into the form for one of the bridge piers and that his remains are entombed in the concrete.

If any roadway that carries you over water can be considered a bridge, the winter ice road that begins in Yellowknife and heads north over 600 kilometres of otherwise impassable territory is one of Canada's longest. It is also fast becoming Canada's most famous span. Open for up to ten of the coldest weeks of the year, the bridge has become known to world-wide audiences through the reality series *Ice Road Truckers*. The program has made minor celebrities out of some of the drivers who, riding on a one hundred centimetre–thick sheet of ice, haul huge loads to the diamond mines north of Yellowknife over frozen sub-Arctic lakes from late January or early February through late March or early April.

I passed through Yellowknife in August. The workday was over for most people as I stood at the corner of 48th Street and 49th Avenue, looking southeast into the long shadows of an August evening (*Corner of 48th Street and 49th Avenue*, p. 163). The streets were all but empty. The downtown looked skeletal, stripped to its bones: a collection of tight curves, hard edges, and angles, as if sketched by an engineer on the back of an envelope. The austerity, it seemed to me, reflected the origins of the place in the 1930s: underground, hard-rock mining, mostly for gold. What is now a city of 20,000 was settled by a pragmatic bunch of prospectors, miners, geologists, and engineers. Find the ore, build the mine, and blast and dig the rock out of the ground; and as for a place to live, well, we may not stay long, so let's just make the best of it on this barren plain of Precambrian rock scraped clean by the glaciers of the last ice age.

I felt the same discomfiting sense of emptiness and transience in other cities (*Mountain Road*, p. 169; *Imperial Hotel*, p. 158). All across the country I saw variants of Yellowknife's unadorned, boxy geometry (*Saskatchewan Drive and Albert Street*, p. 179; *Brant Hotel*, p. 187; *Winnipeg Art Gallery*, p. 174; *Cityview Restaurant*, p. 186; *Wall*, p. 185), an observation that reinforced my (possibly off-the-mark) vision of most Canadian cities as rooted in expediency: founded as temporary, makeshift places geared to the extraction and exploitation of natural resources.

Exceptions prove the rule. In urban districts or neighbourhoods the edges are softer and a sense of history and permanence prevails (*Dufferin Terrace*, p. 188; *Christie Barber Shop*, p. 175; *Homer Café*, p. 182; *St. John's Harbour*, p. 189). Whitehorse (*Yukon Theatre*, p. 173), too, has a delicateness that rounds out its rough edges and belies its origins as a remote transportation hub serving the Klondike gold rush at the end of the nineteenth century.

Only rarely, I think, will you find anything approaching genuine grandeur in a Canadian city. And when you do, it's the presence of the natural environment, hovering just outside the city, that's responsible. Visit Vancouver and your eye is drawn to the sea and the mountains. (The line of the overhead roadway cutting diagonally across the frame in *Lions Gate*, p. 160, pulls your eye right into the mountains, so much so that you hardly notice the presence of West Vancouver at their base). In Quebec City, as you stand on the broad terrace in front of the Chateau Frontenac hotel (*Dufferin Terrace*, p. 188) and look east toward Lévis, the grandeur of the St. Lawrence River dwarfs everything else. Jacques Cartier, the French explorer who arrived in what is now Quebec City (then, the Aboriginal settlement of Stadacona) in 1535, thought the river was the much-sought-after road to the riches of China.

Cartier sailed under a commission from King Francis I of France, with instructions to "discover certain islands and lands where it is said that a great quantity of gold and other precious things are to be found." And so even here, in one of the country's oldest continuously settled places, you can't help but be reminded that Canada was built on the efforts of people who were just passing through on the road to hoped-for riches.

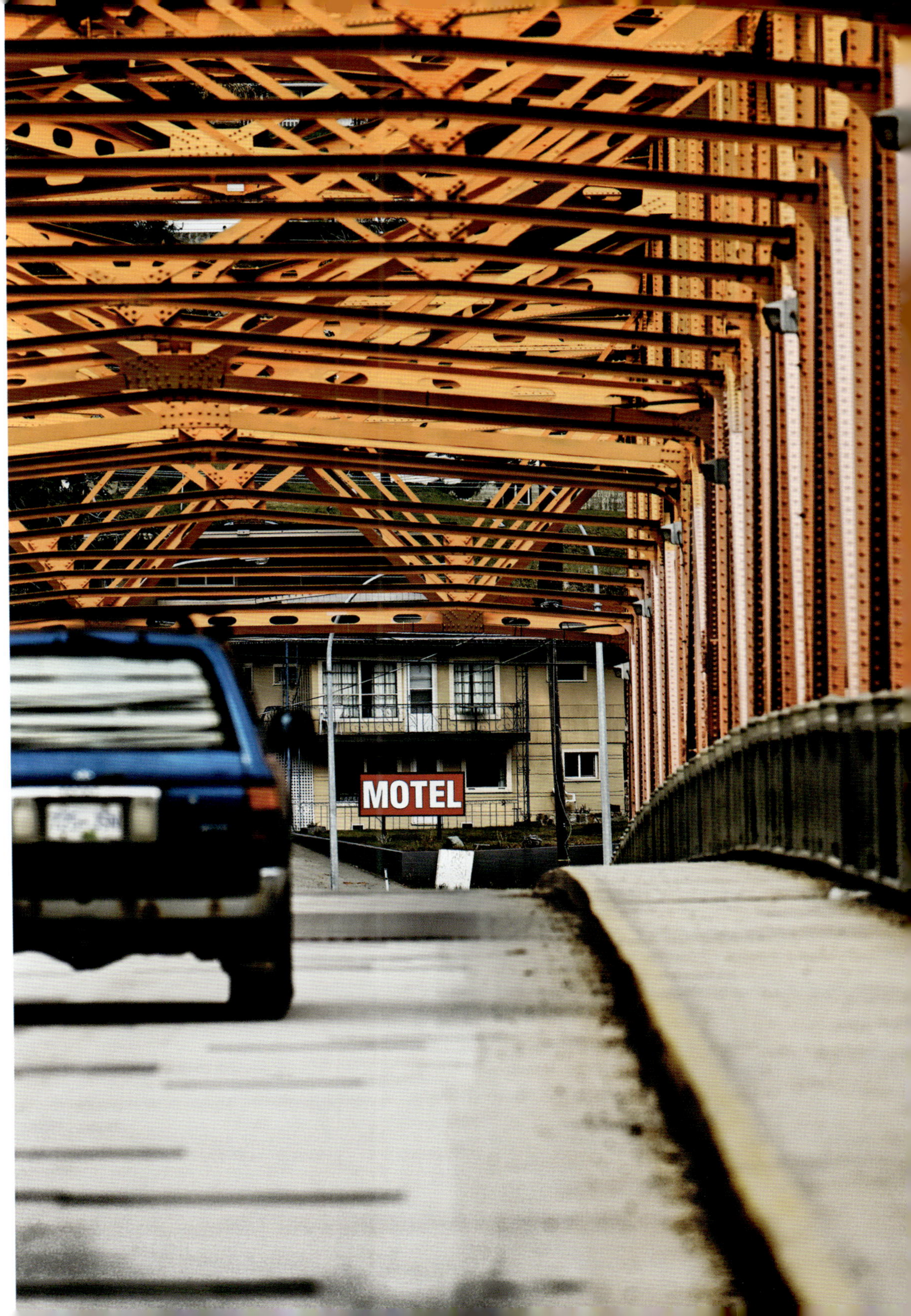

Motel, Nelson BC

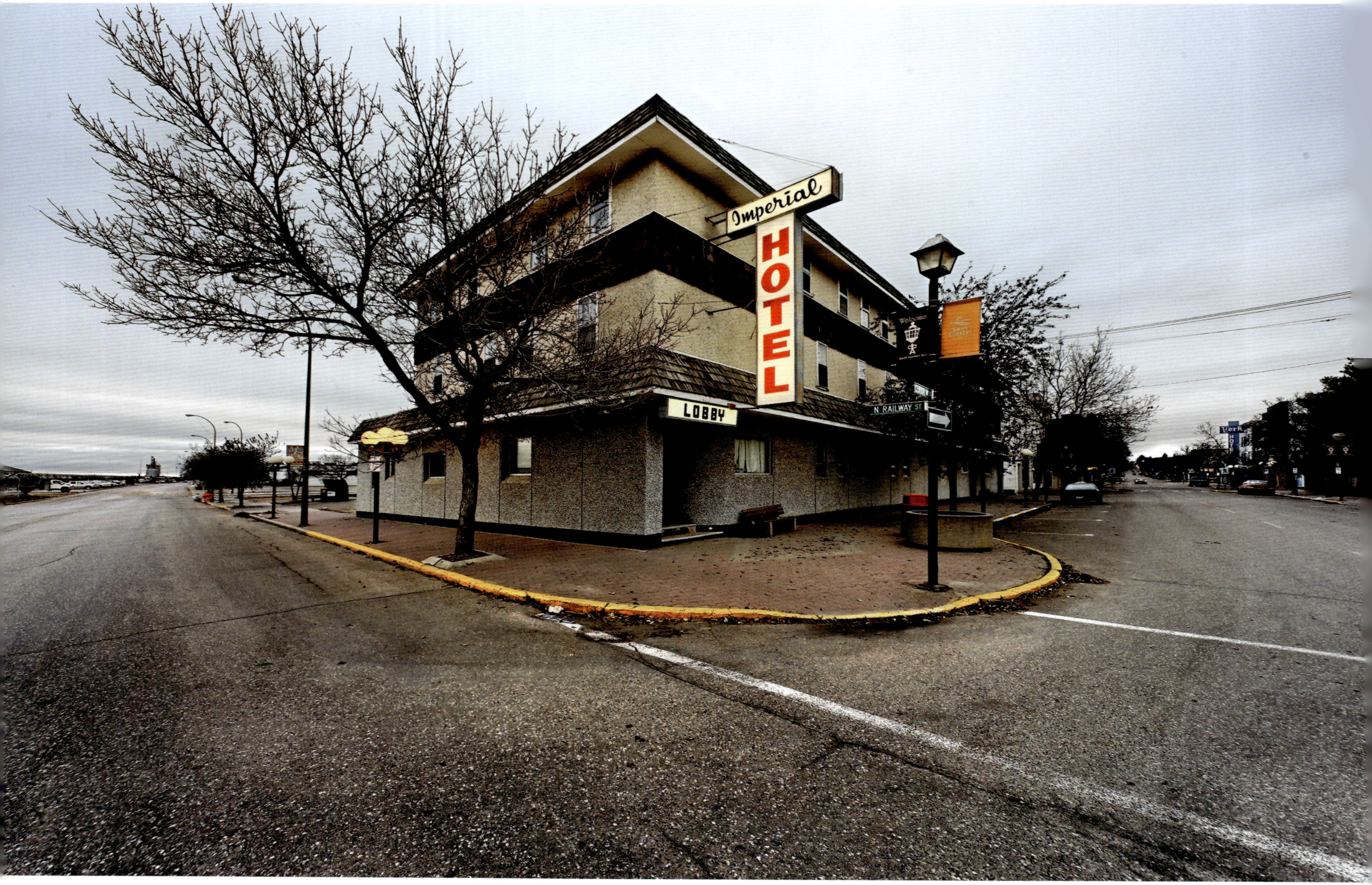

Imperial Hotel, Swift Current SK

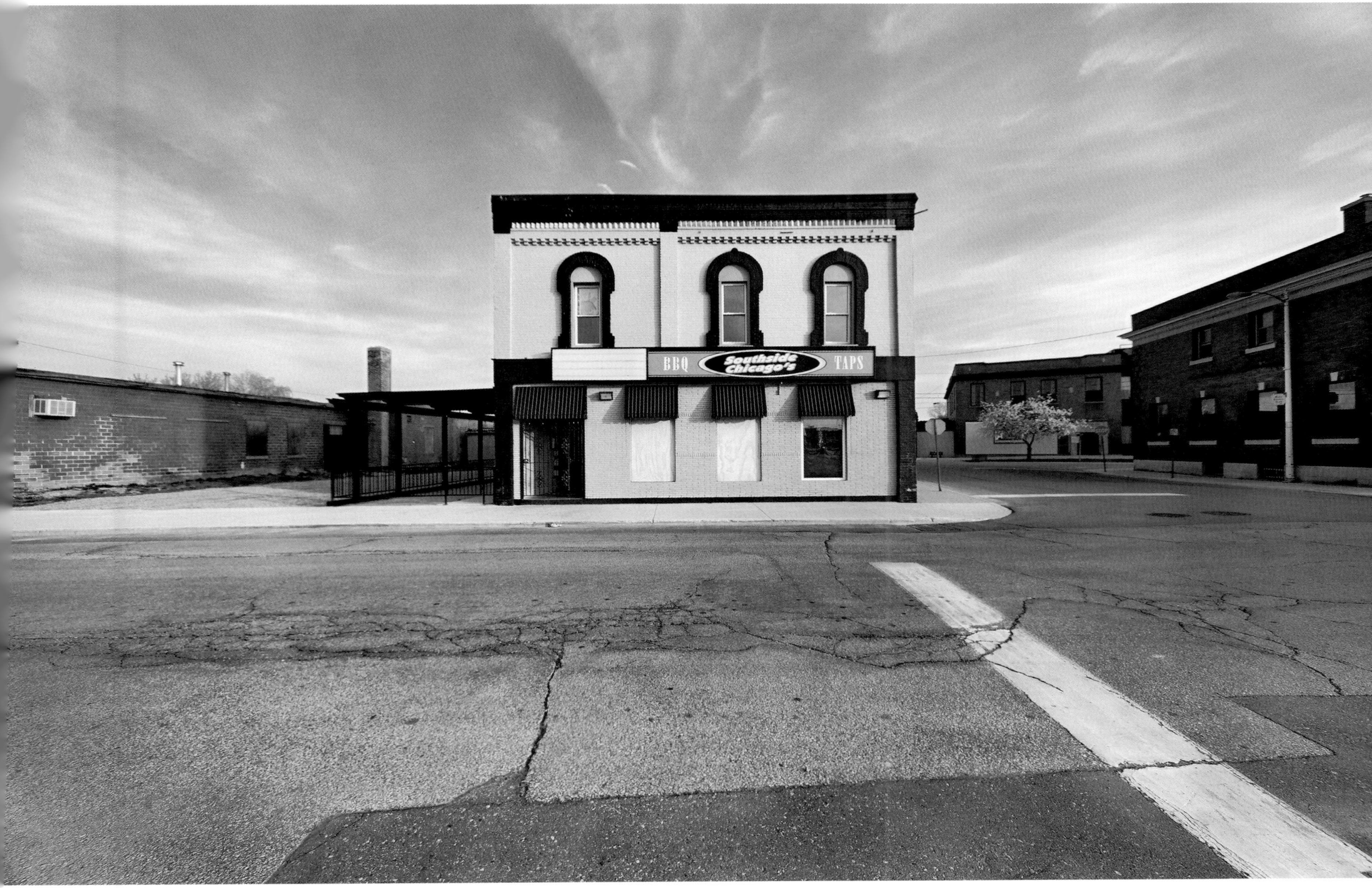

Bar and Grill, Wallaceburg ON

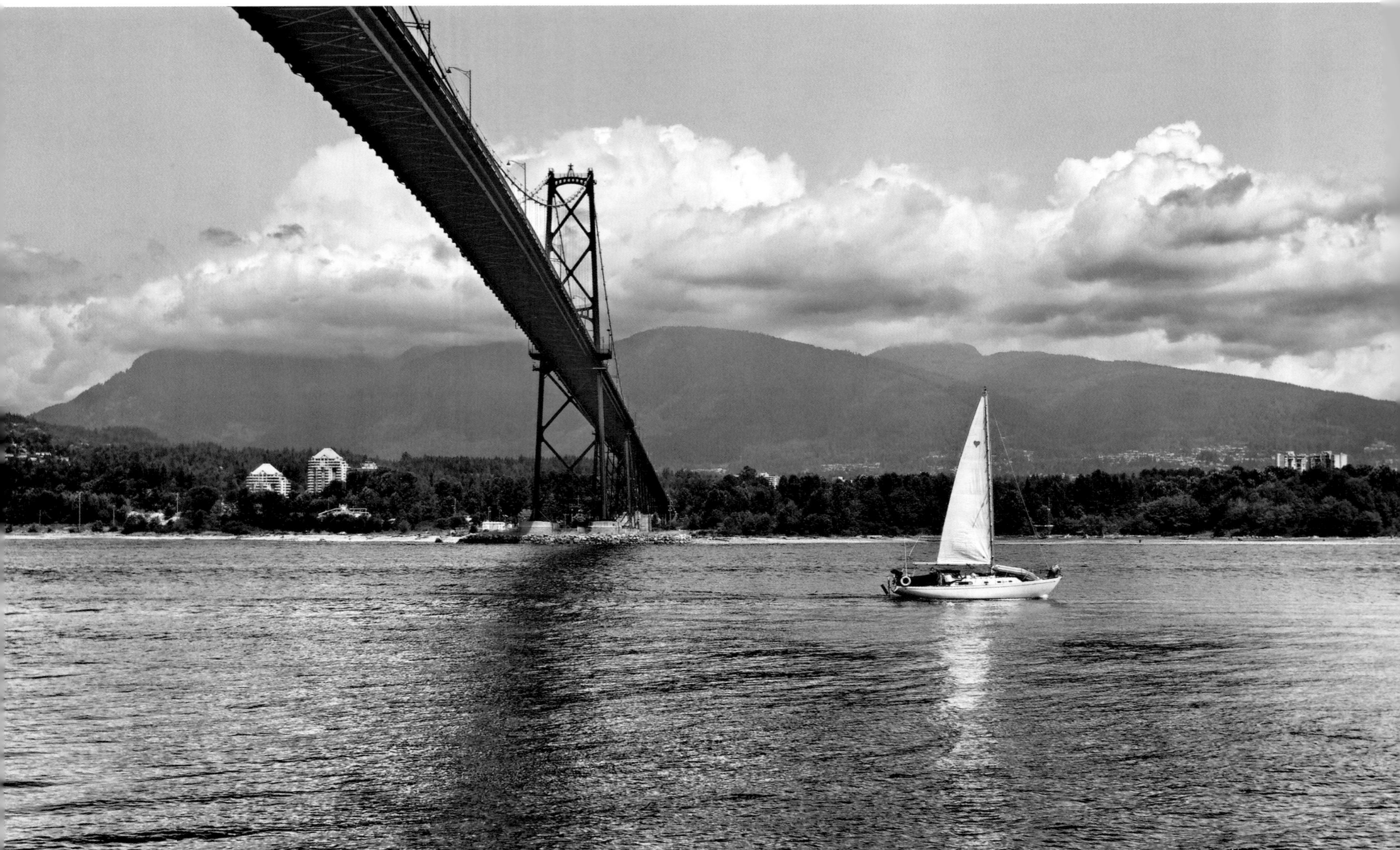

Lions Gate, Vancouver BC

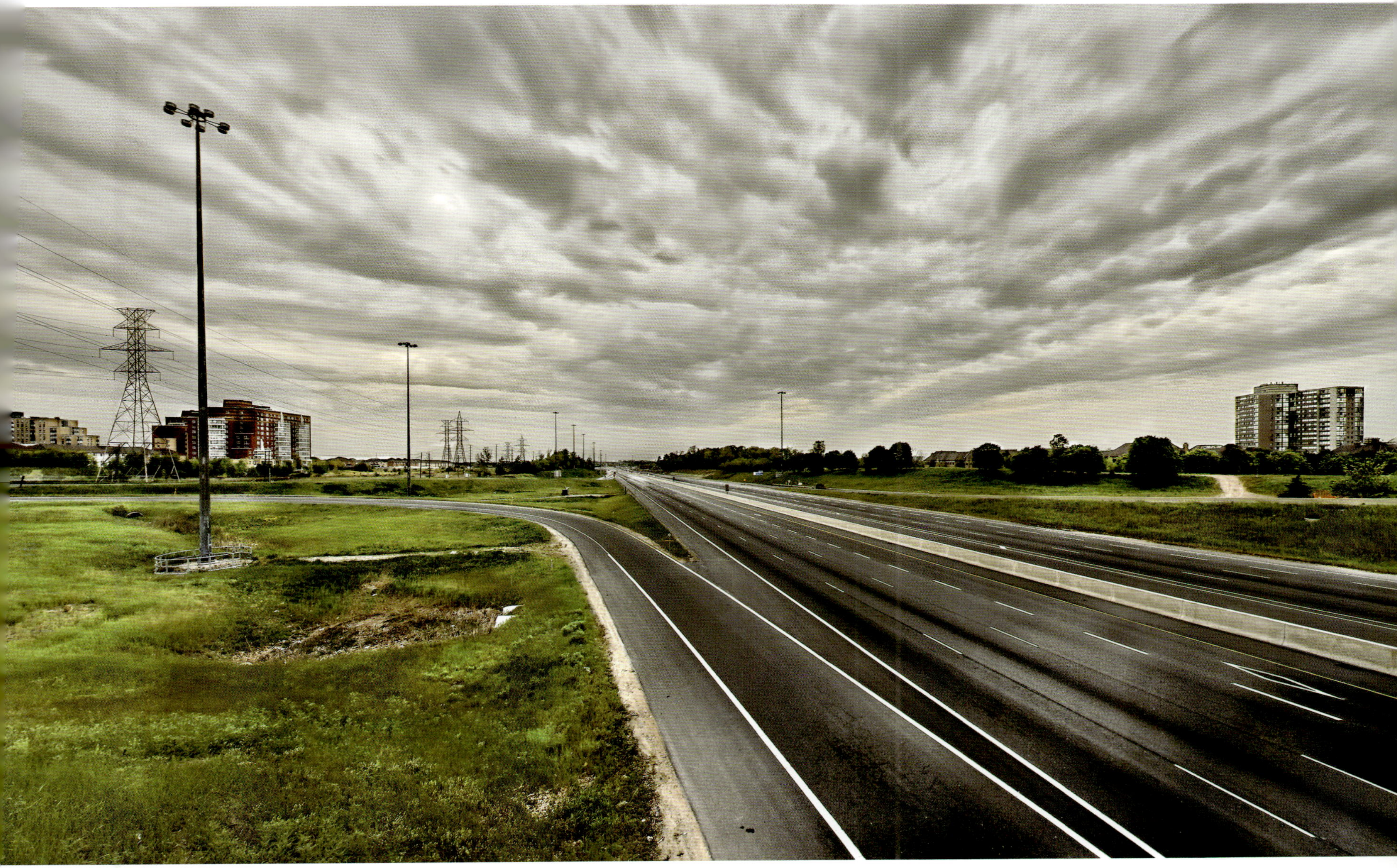

On Ramp, Mississauga ON

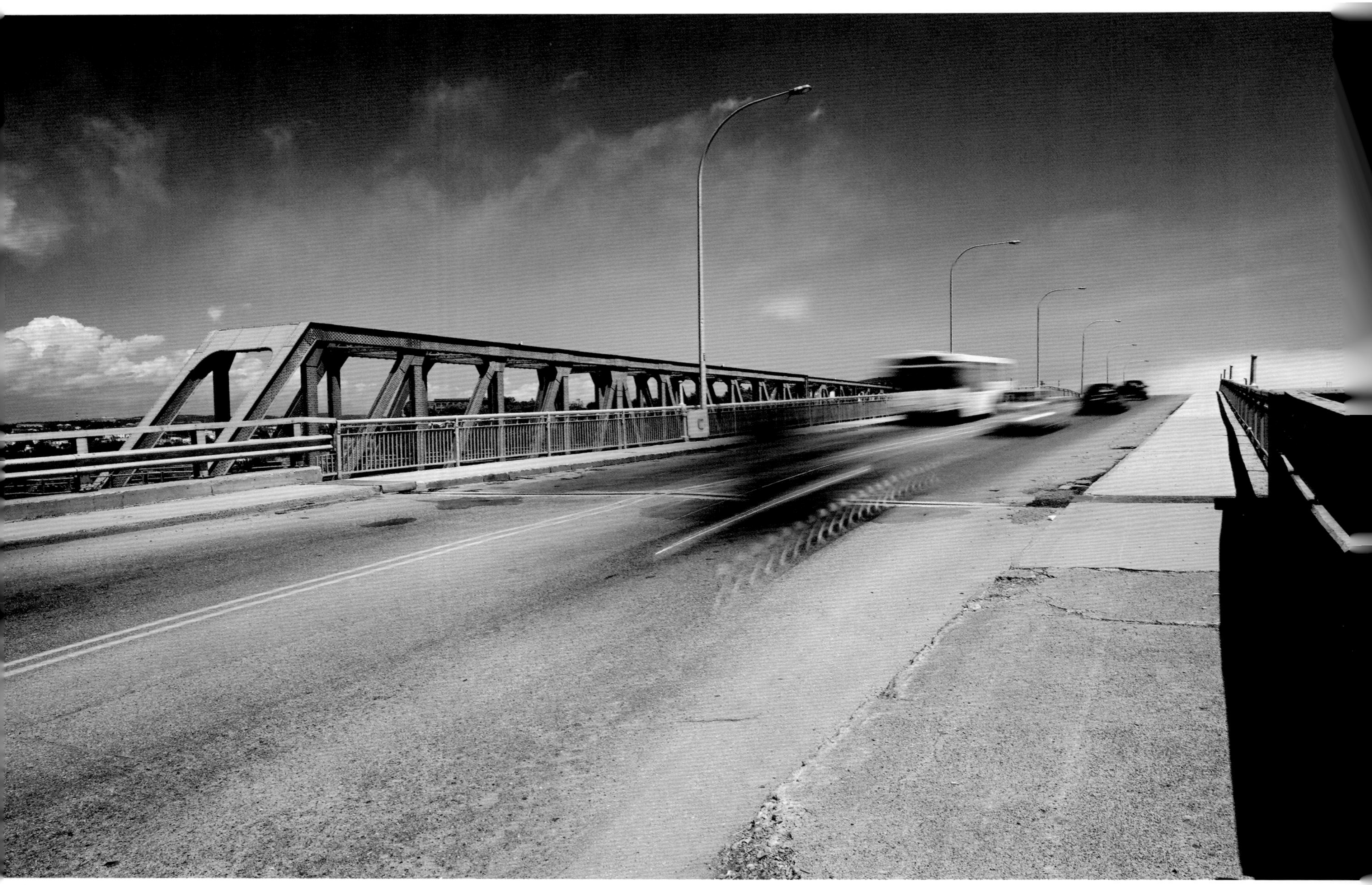

Bridge Street, Saint John NB

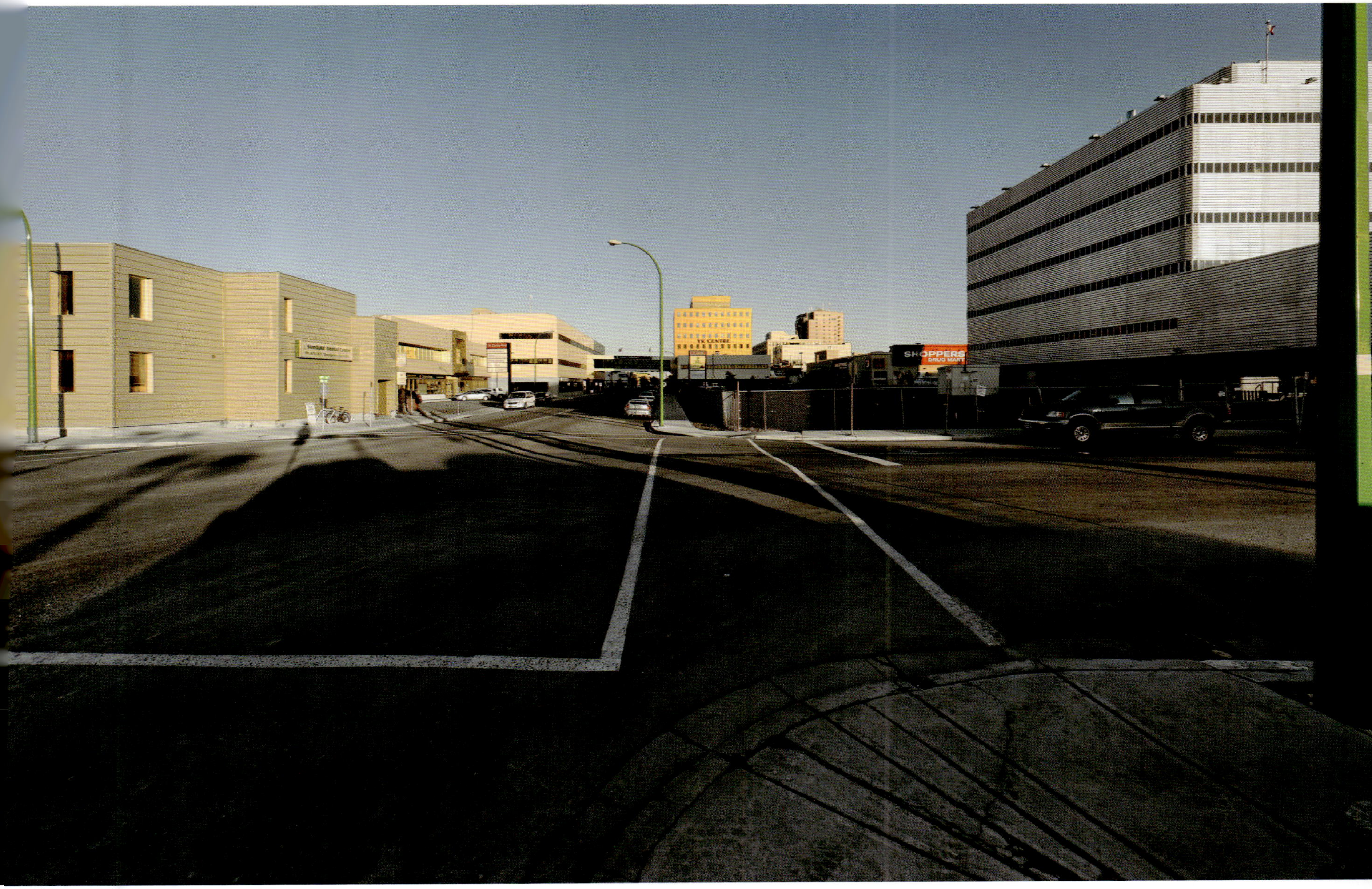

Corner of 48th Street and 49th Avenue, Yellowknife NT

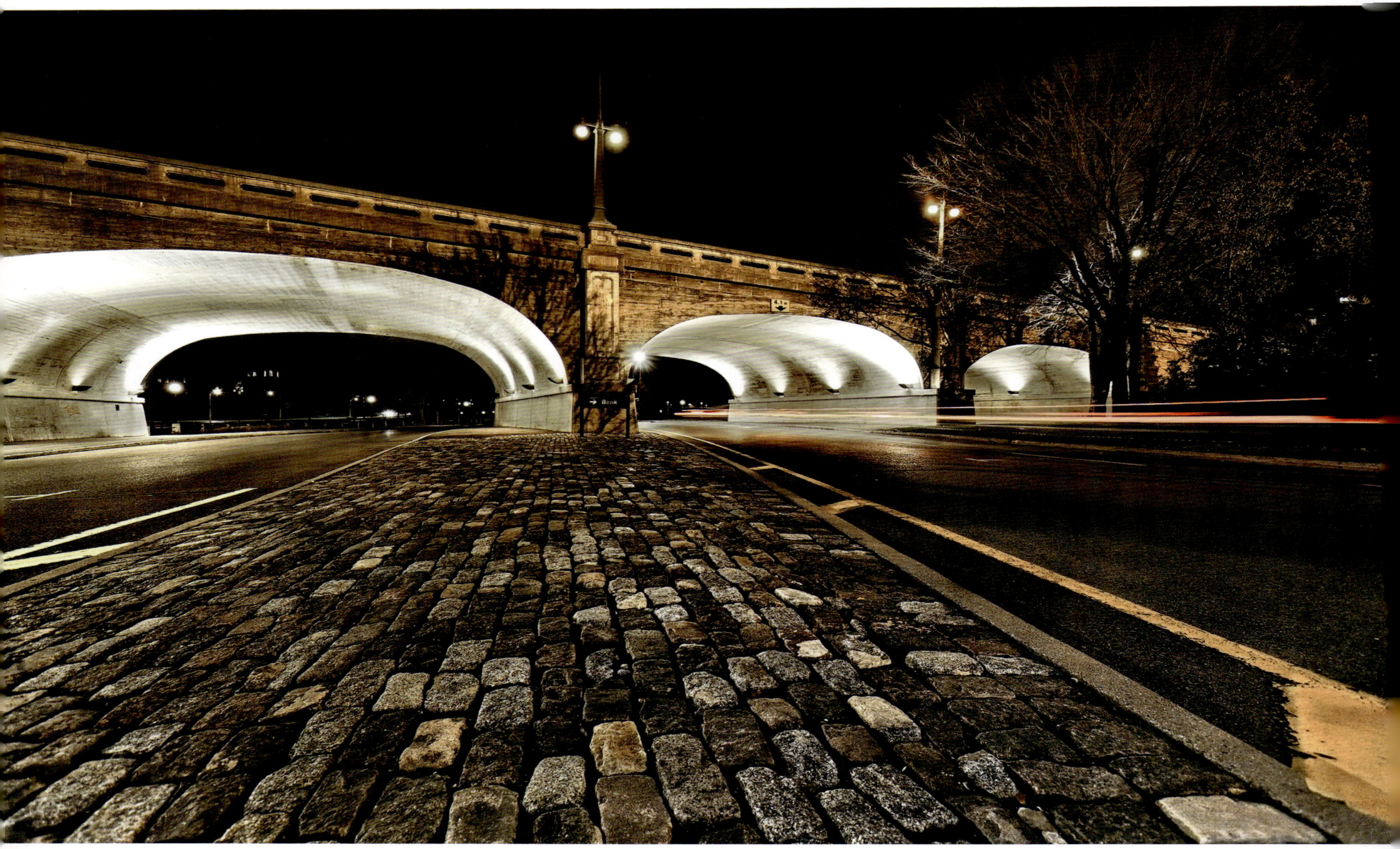

Bank Street Bridge, Ottawa ON

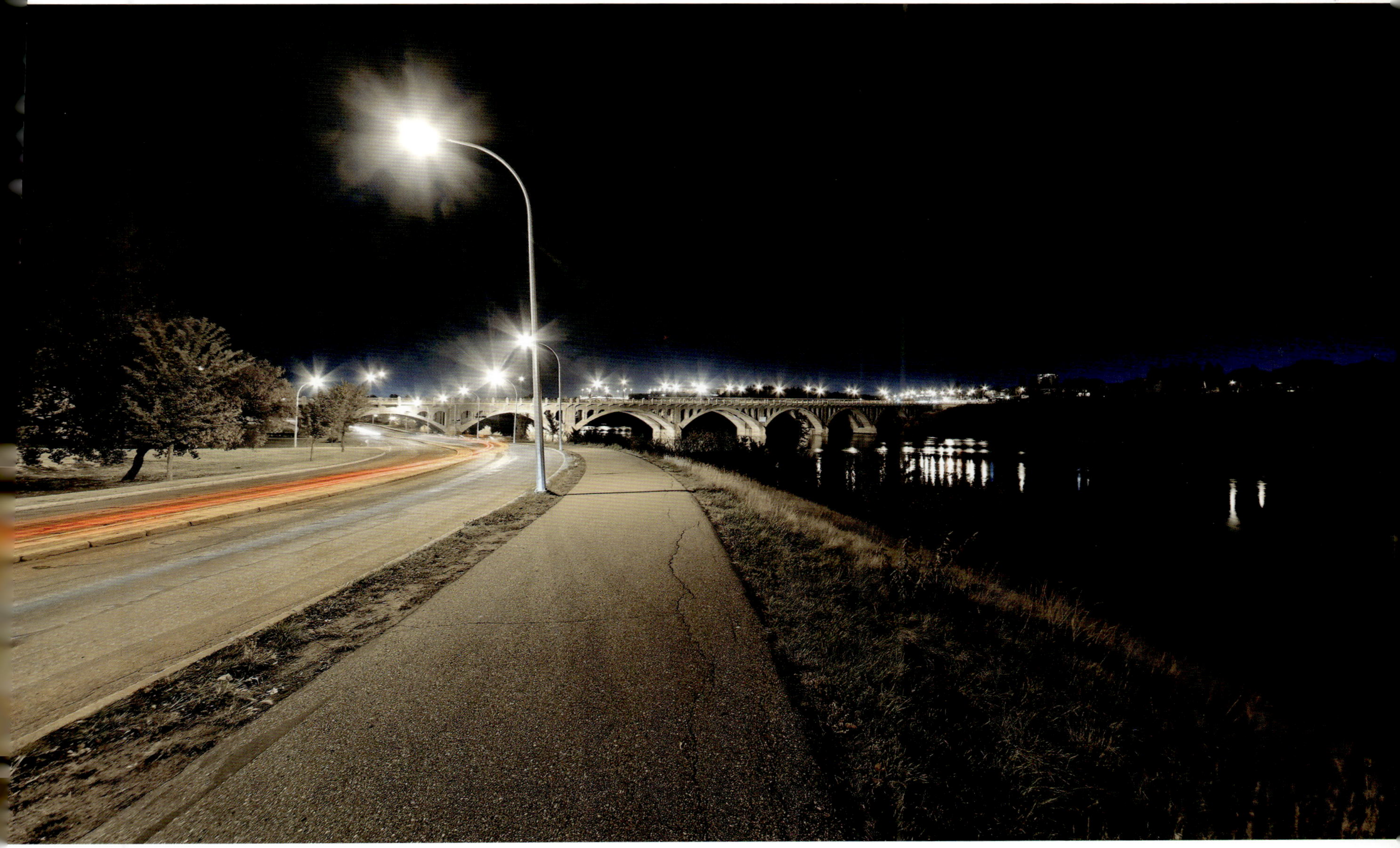

University Bridge, Saskatoon SK

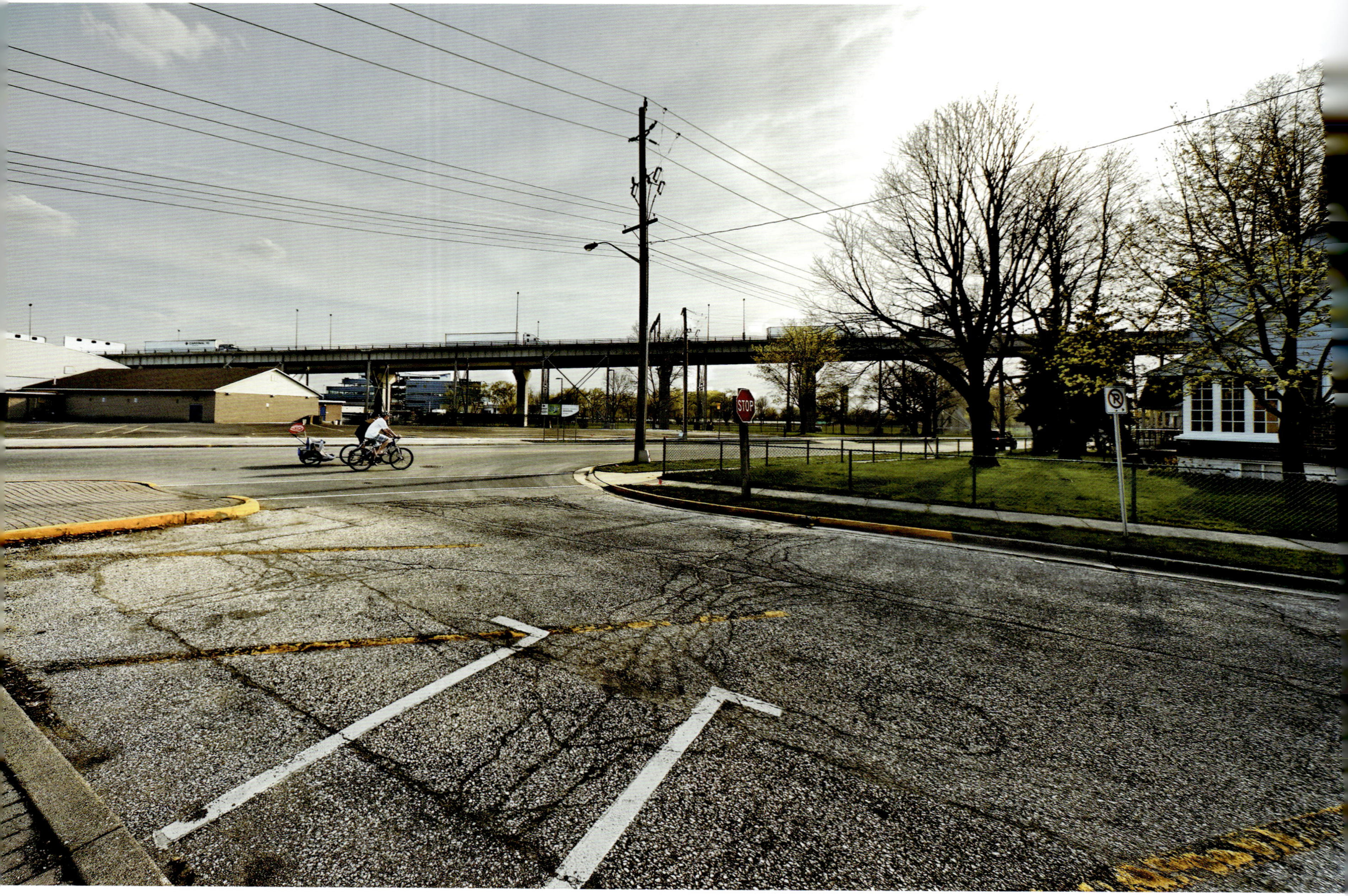

Bridge to USA, Sarnia ON

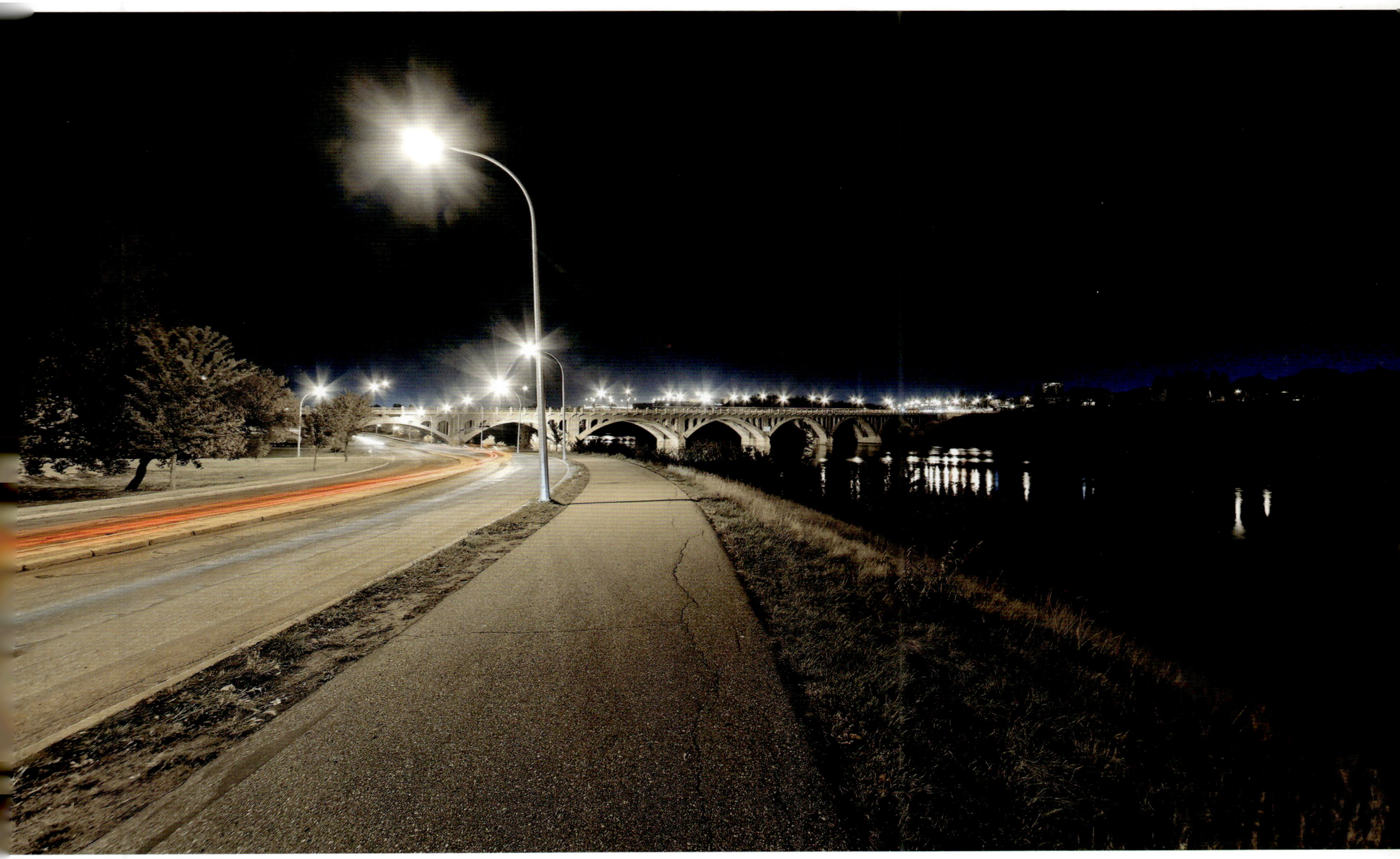

University Bridge, Saskatoon SK

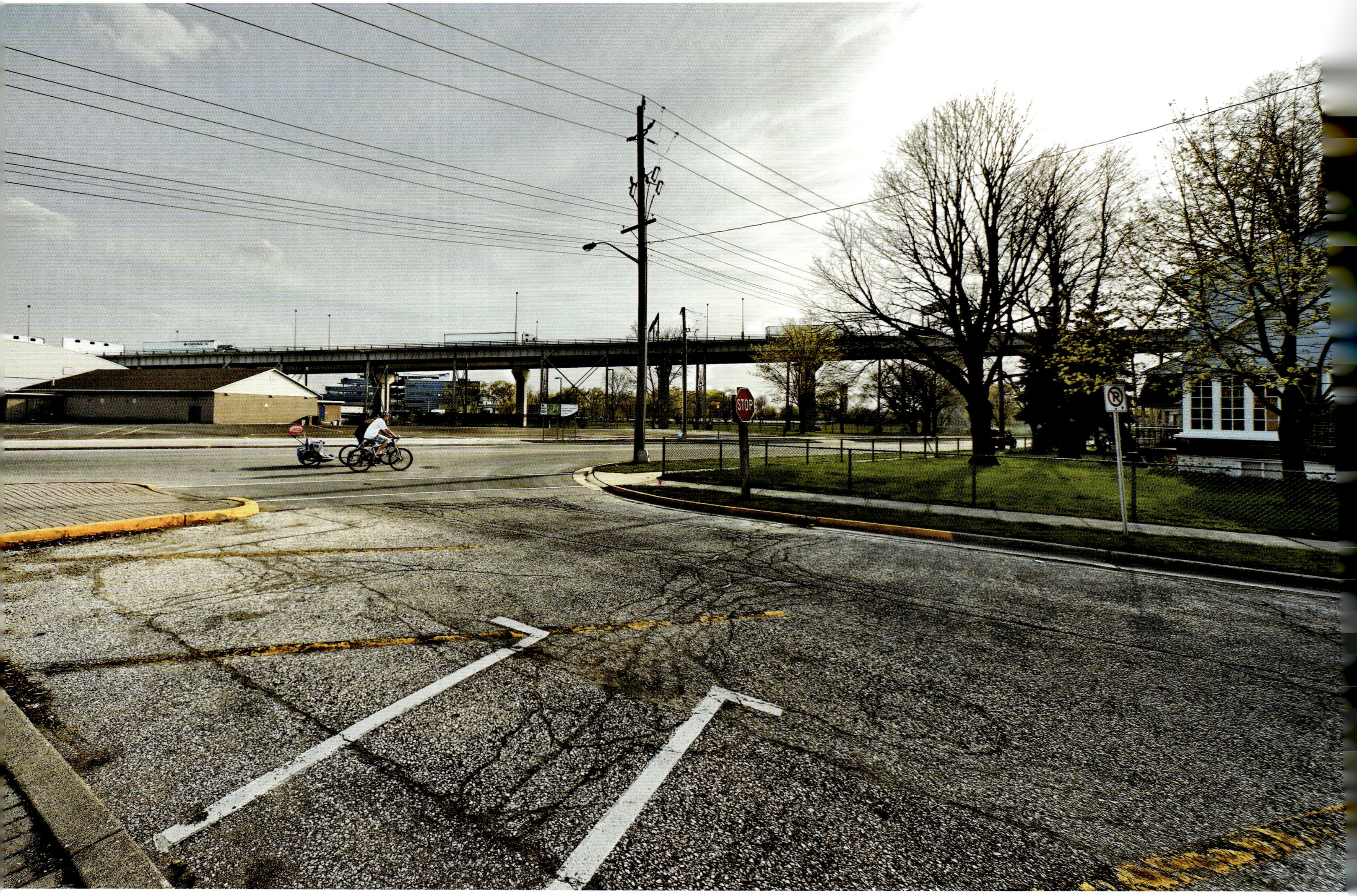

Bridge to USA, Sarnia ON

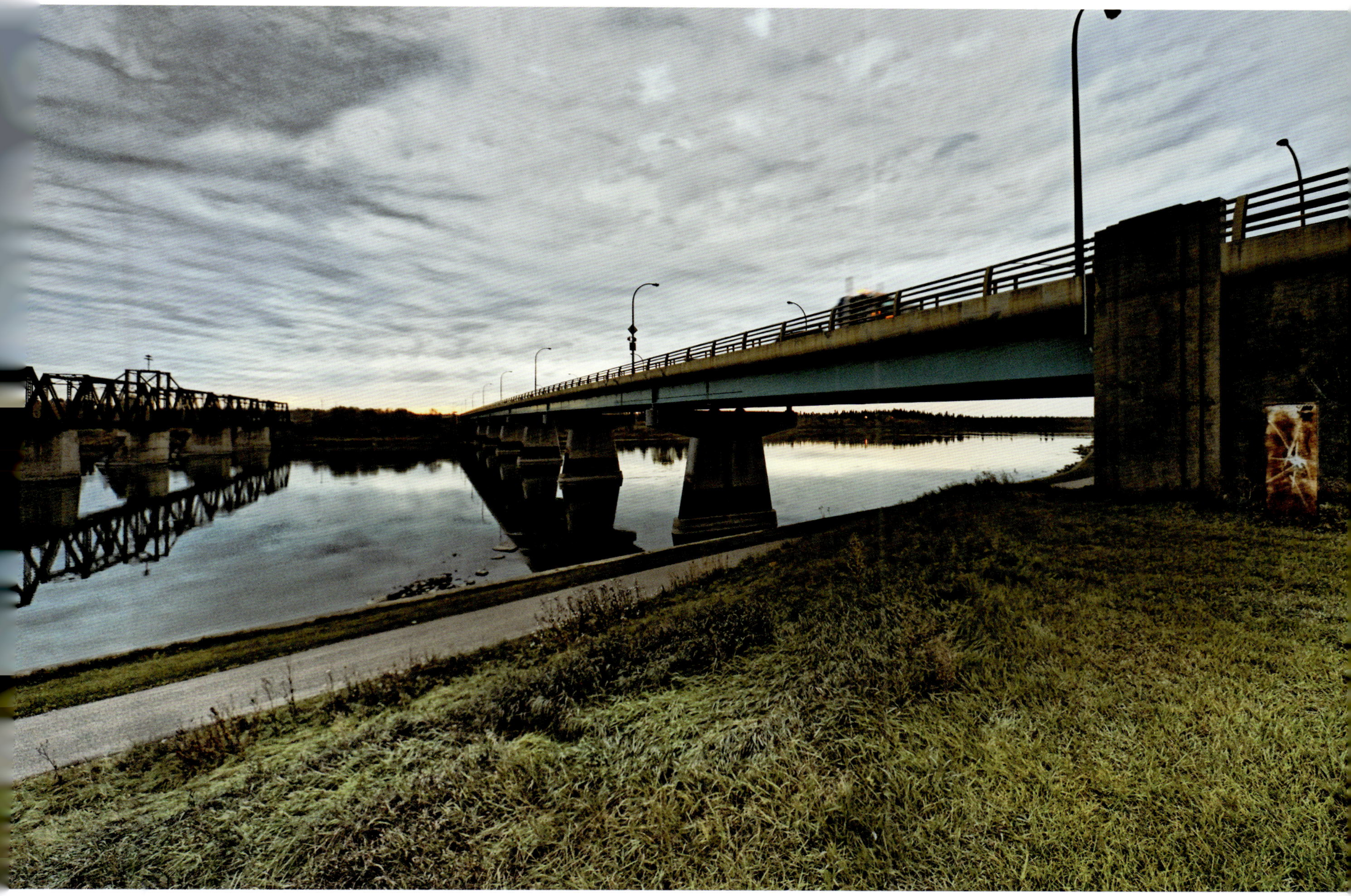

North Saskatchewan River, Prince Albert SK

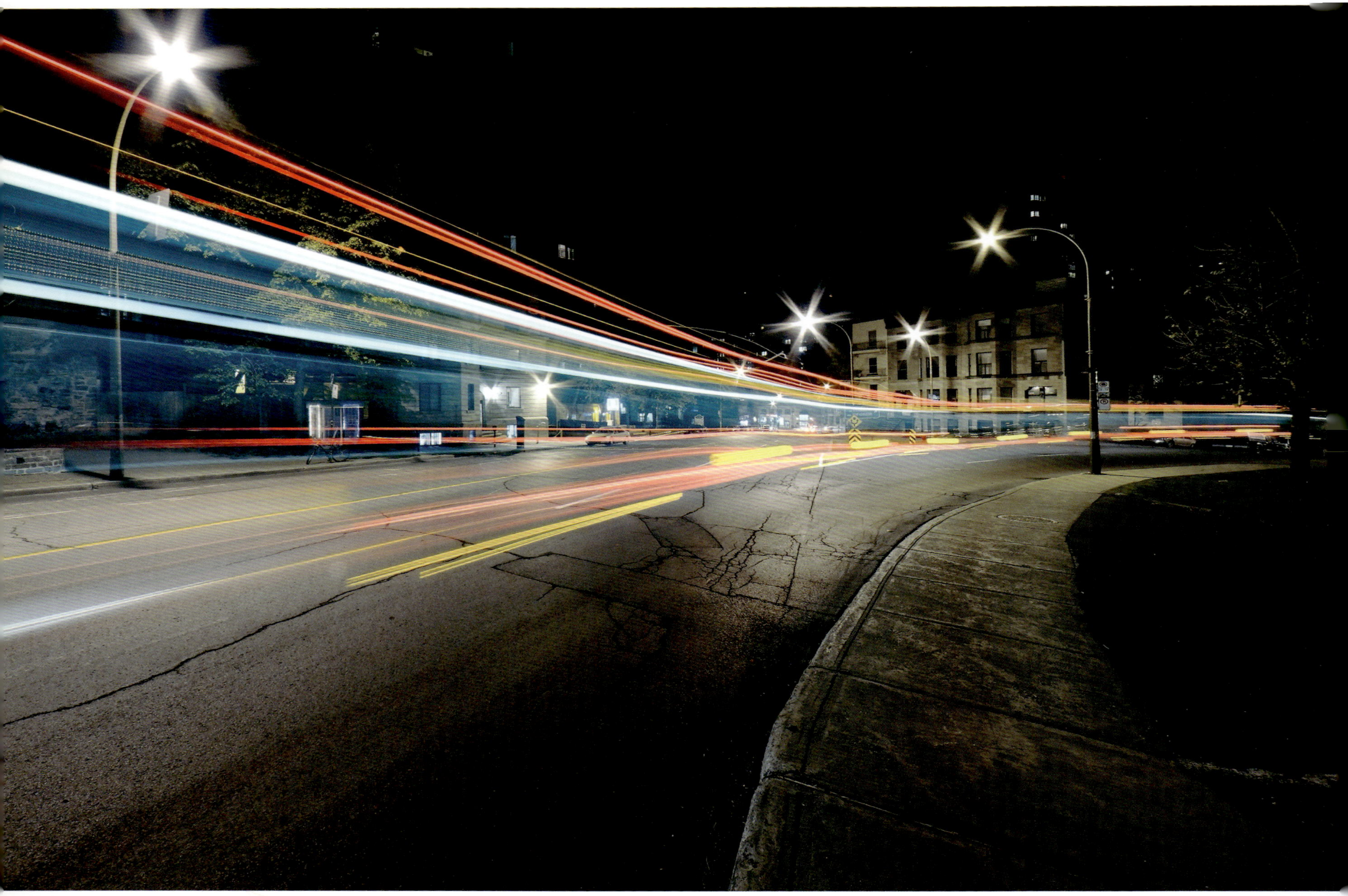

Midnight Bus, Montreal QC

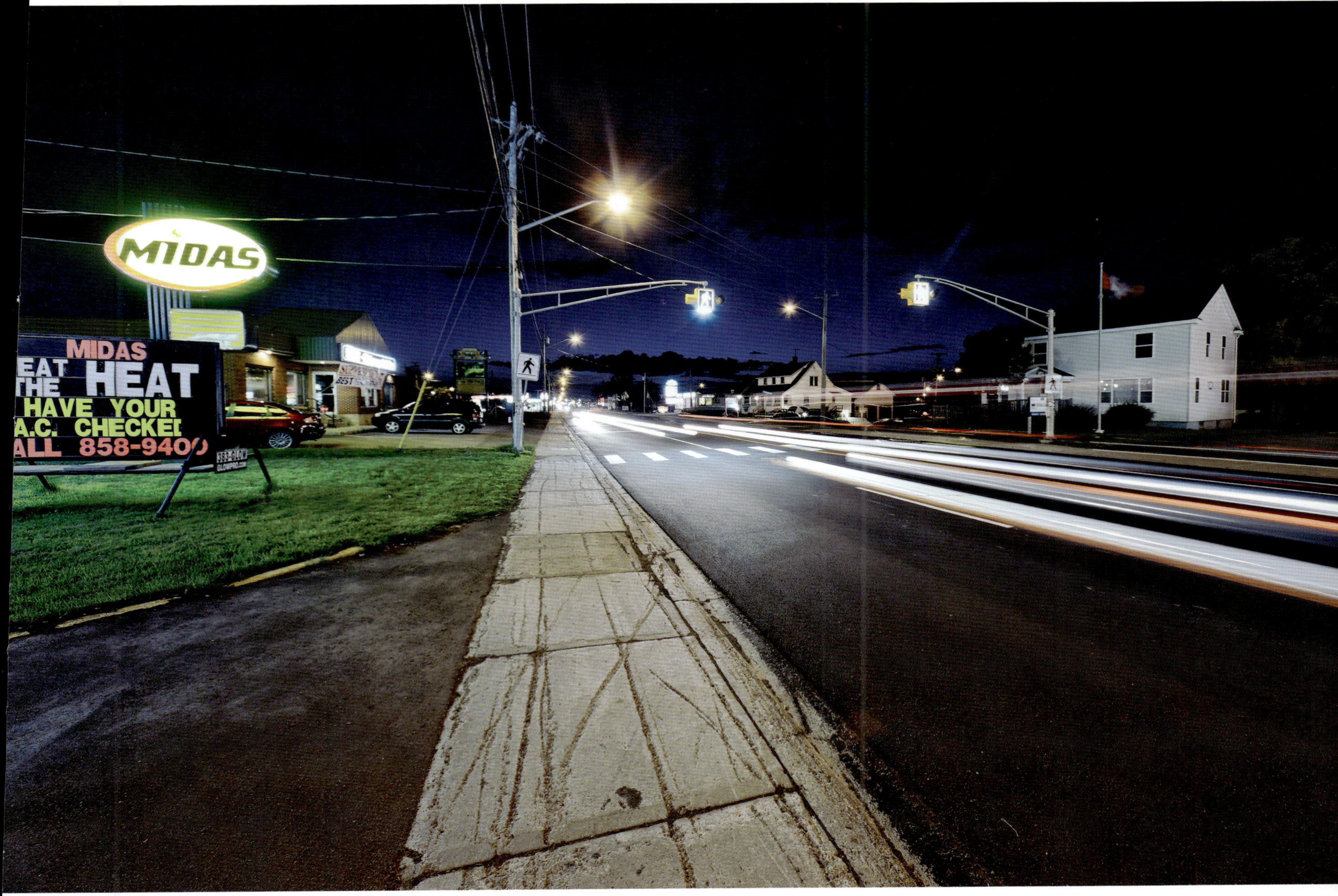

Mountain Road, Moncton NB

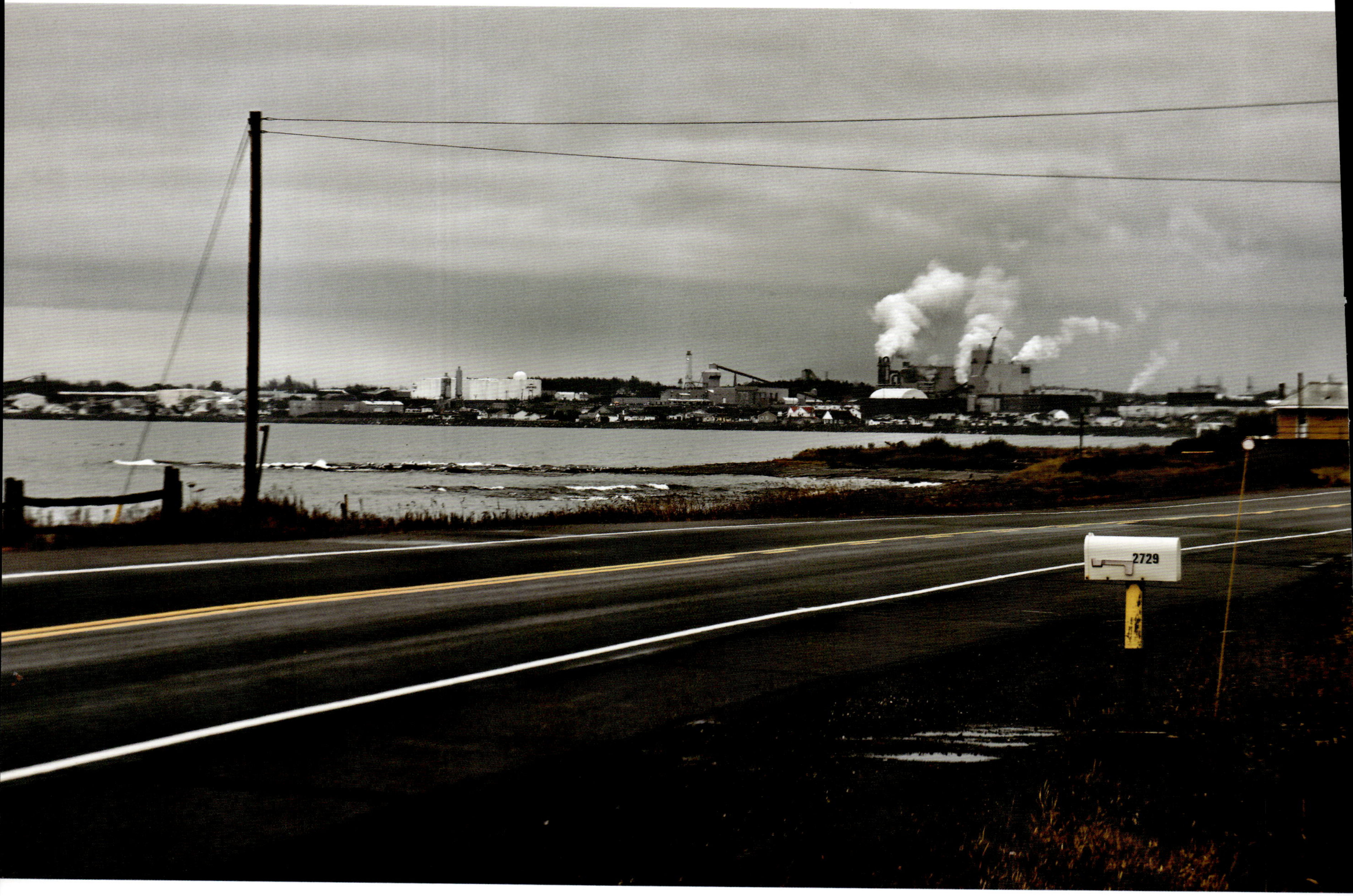

Paper Mill, Matane QC

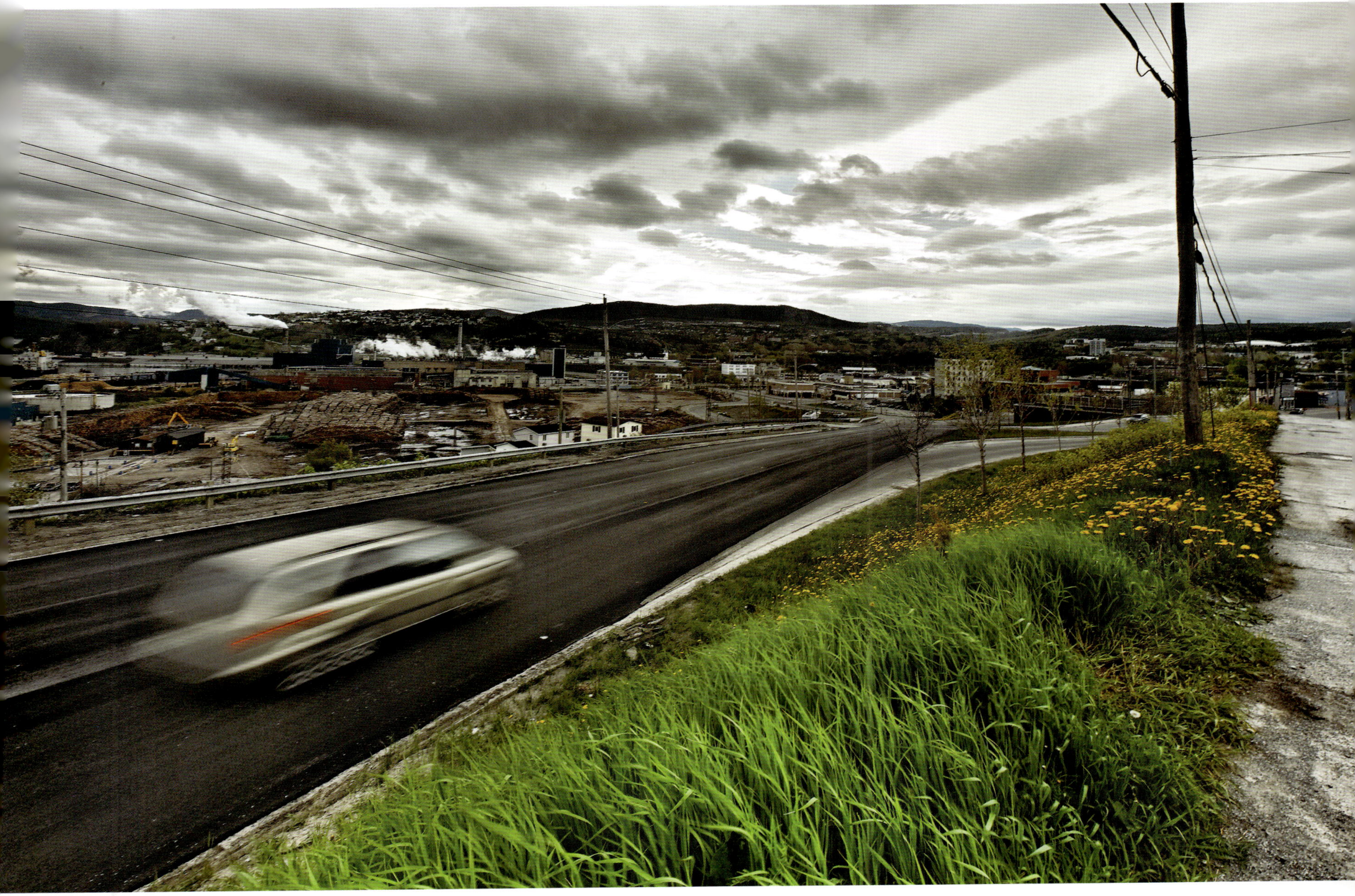

Corner Brook, Corner Brook NL

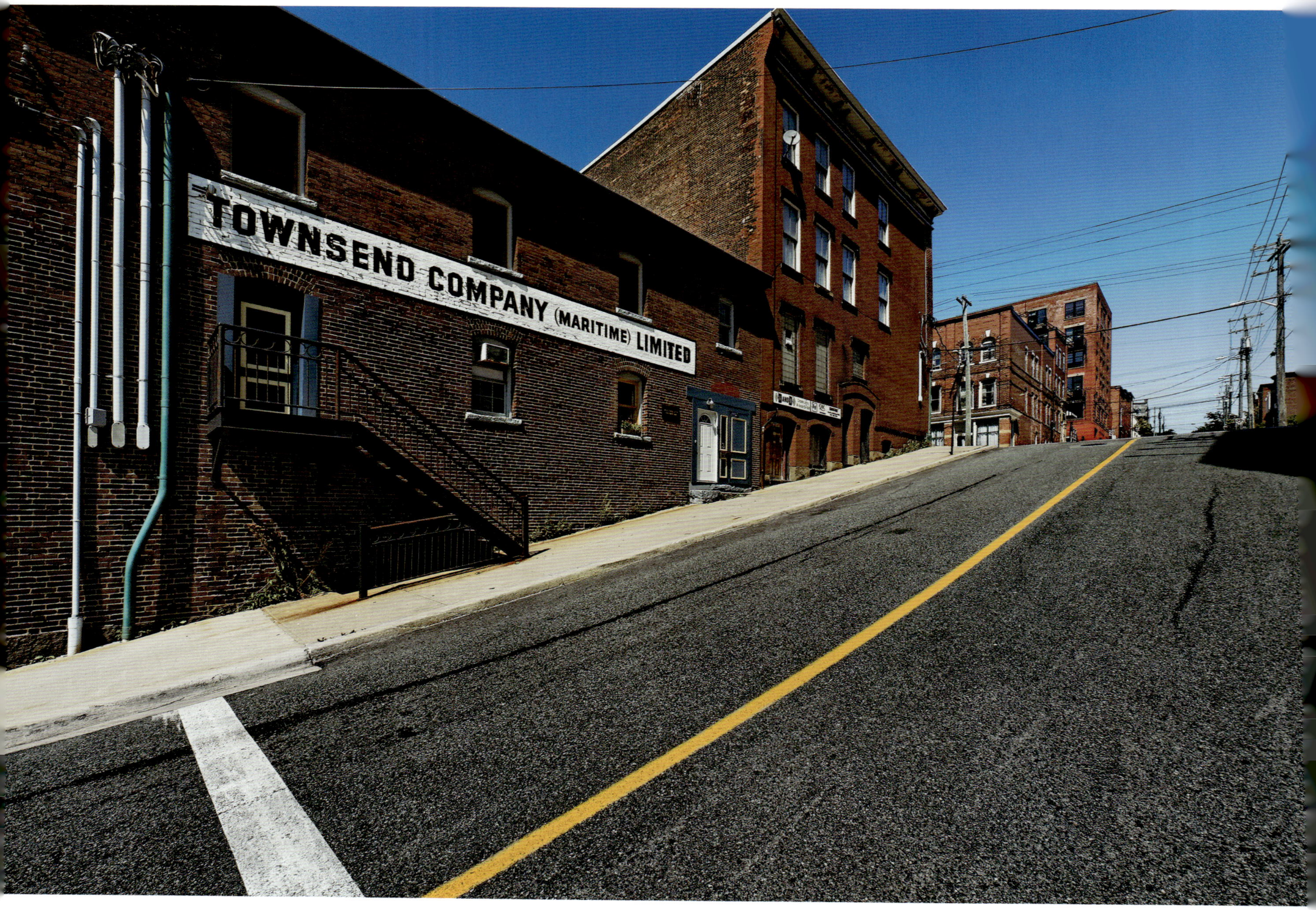

Townsend Building, Saint John NB

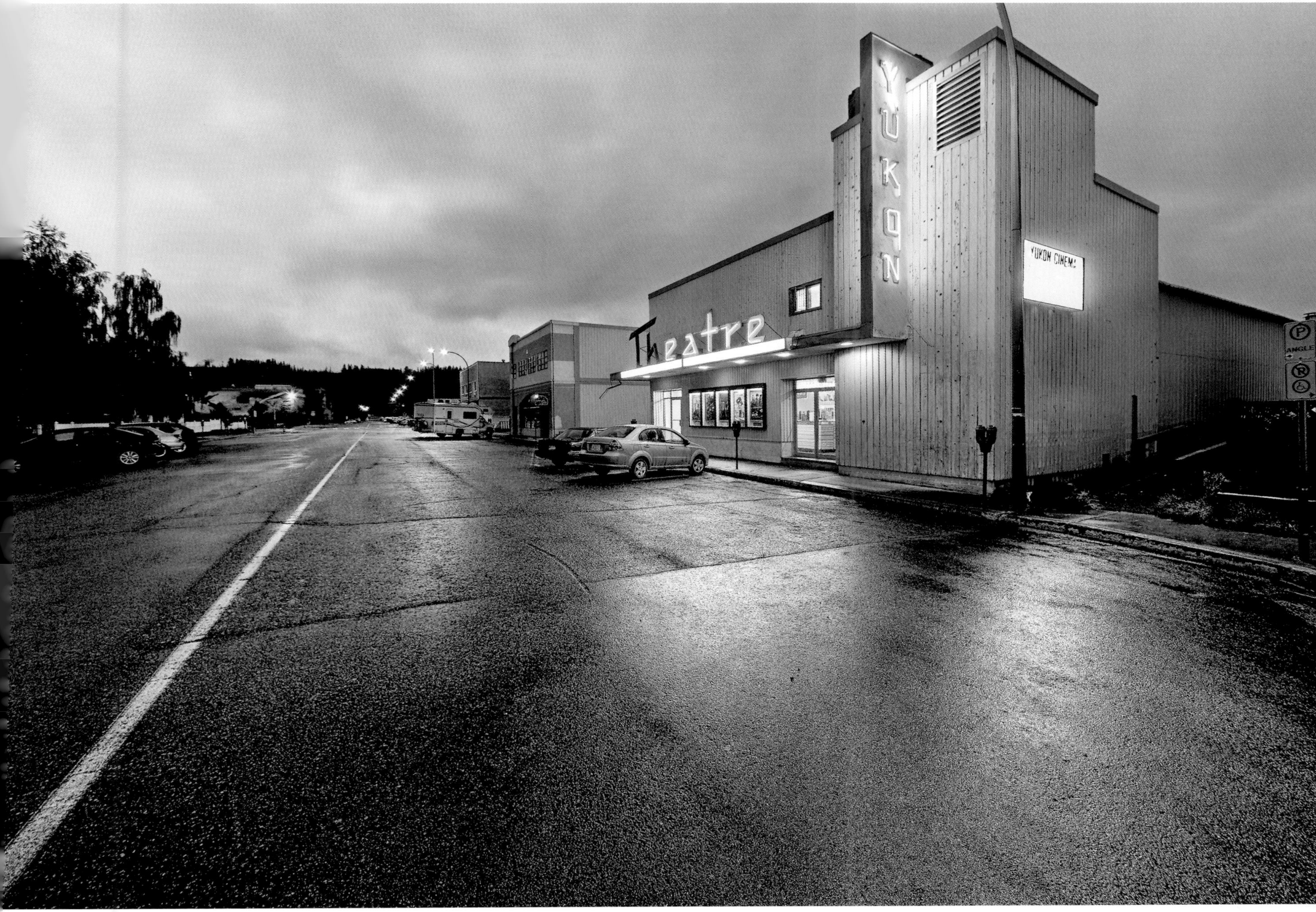

Yukon Theatre, Whitehorse YK

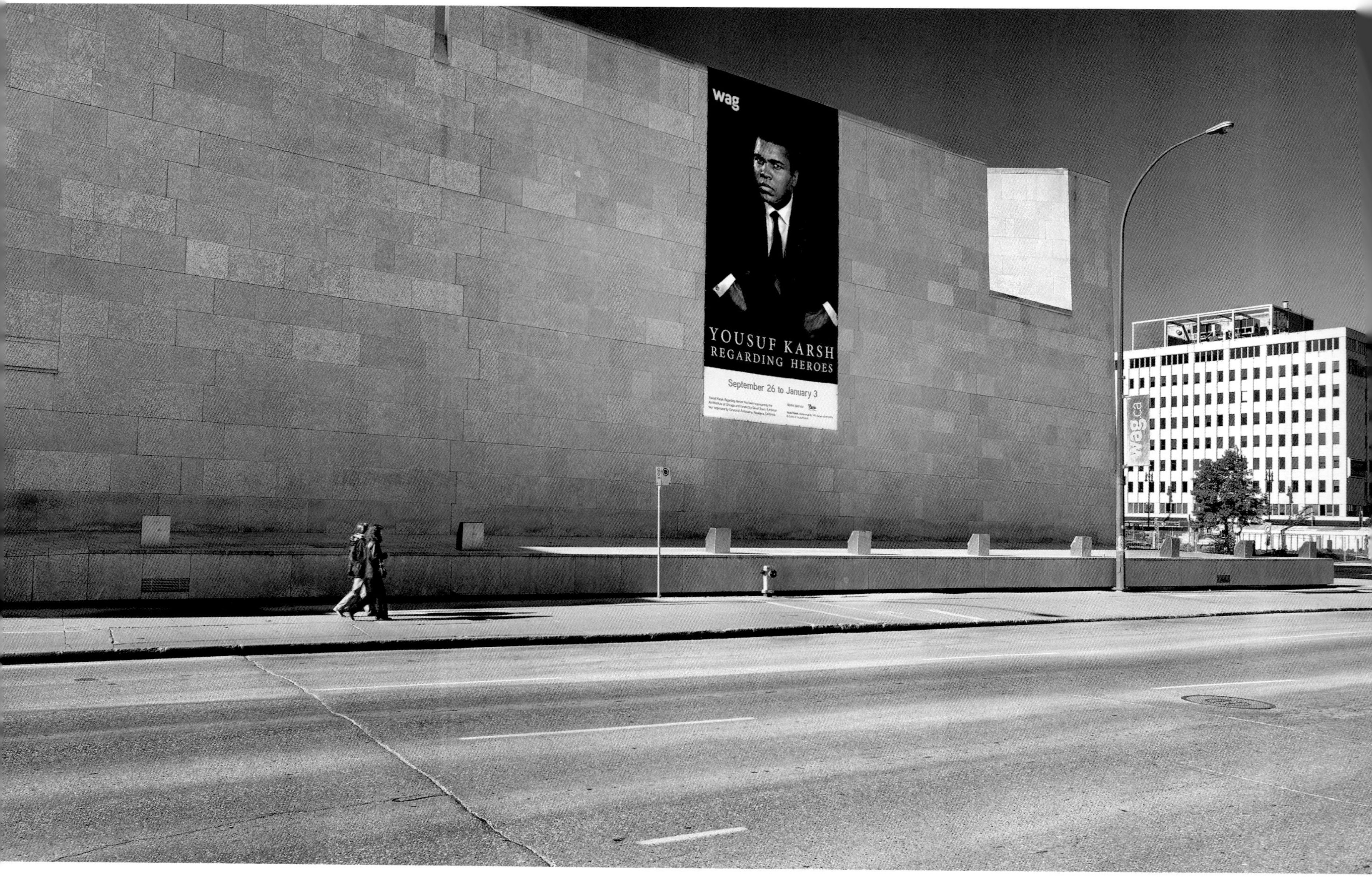

Winnipeg Art Gallery, Winnipeg MB

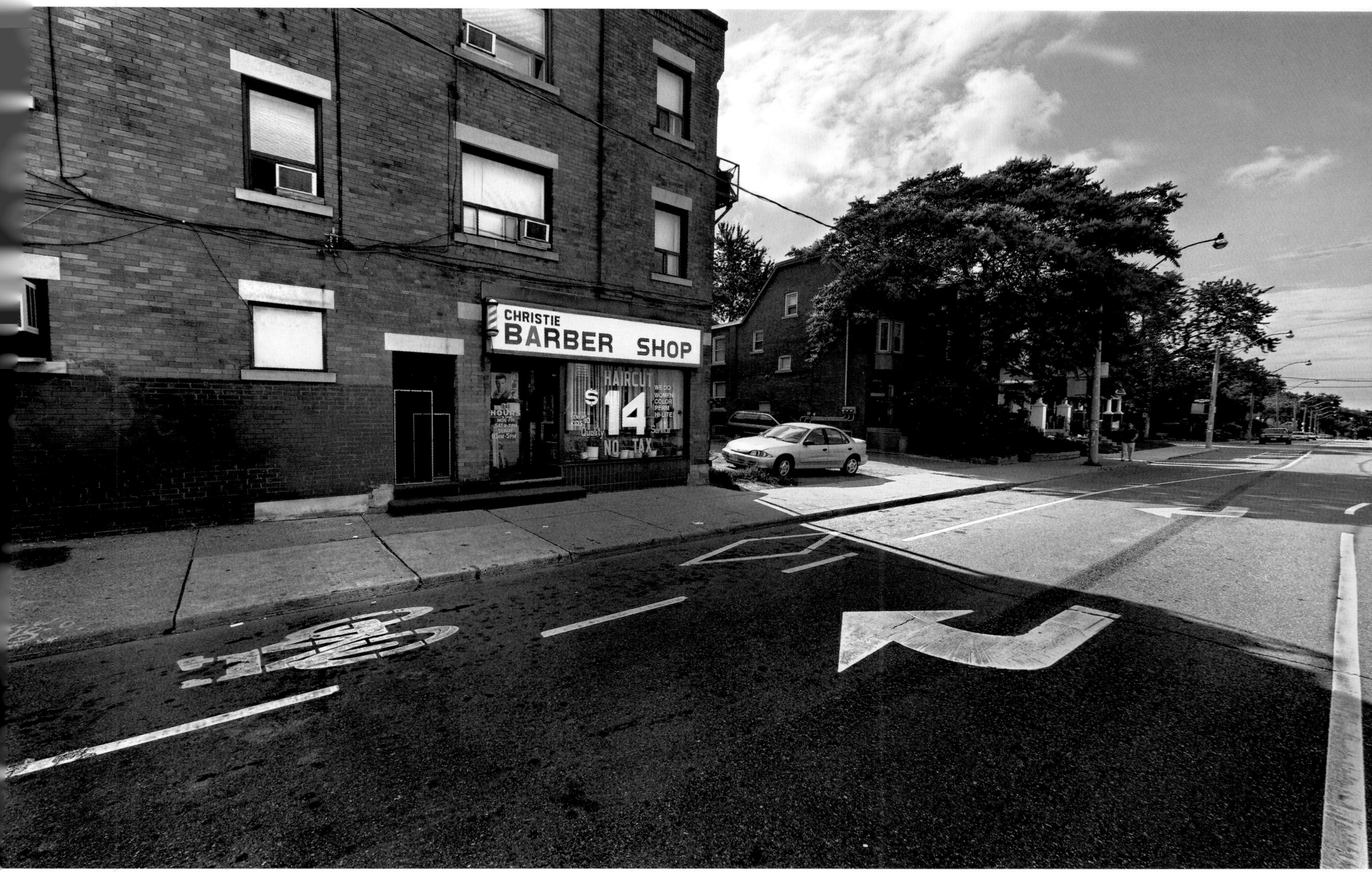

Christie Barber Shop, Toronto ON

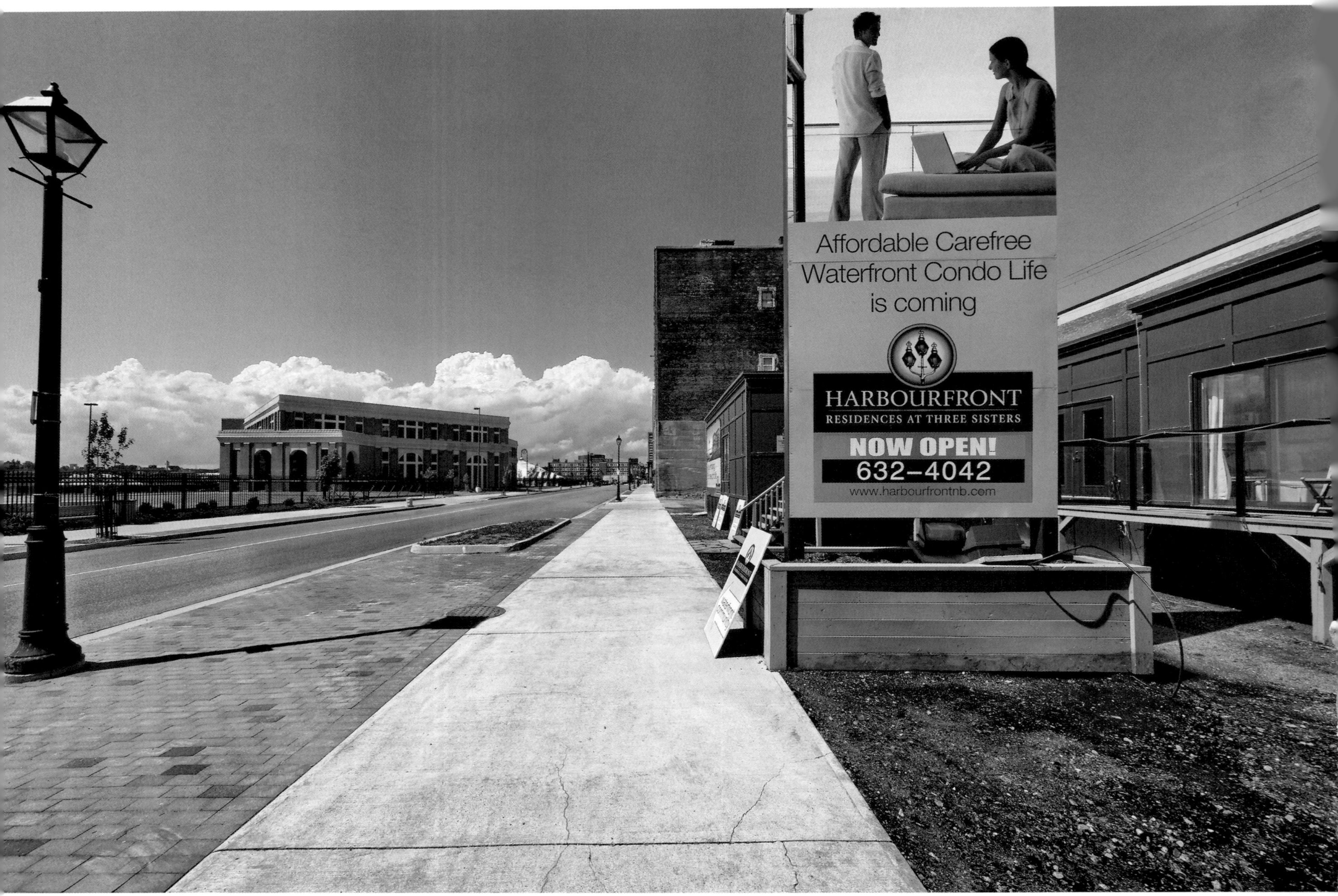

Condo Life is Coming, Saint John NB

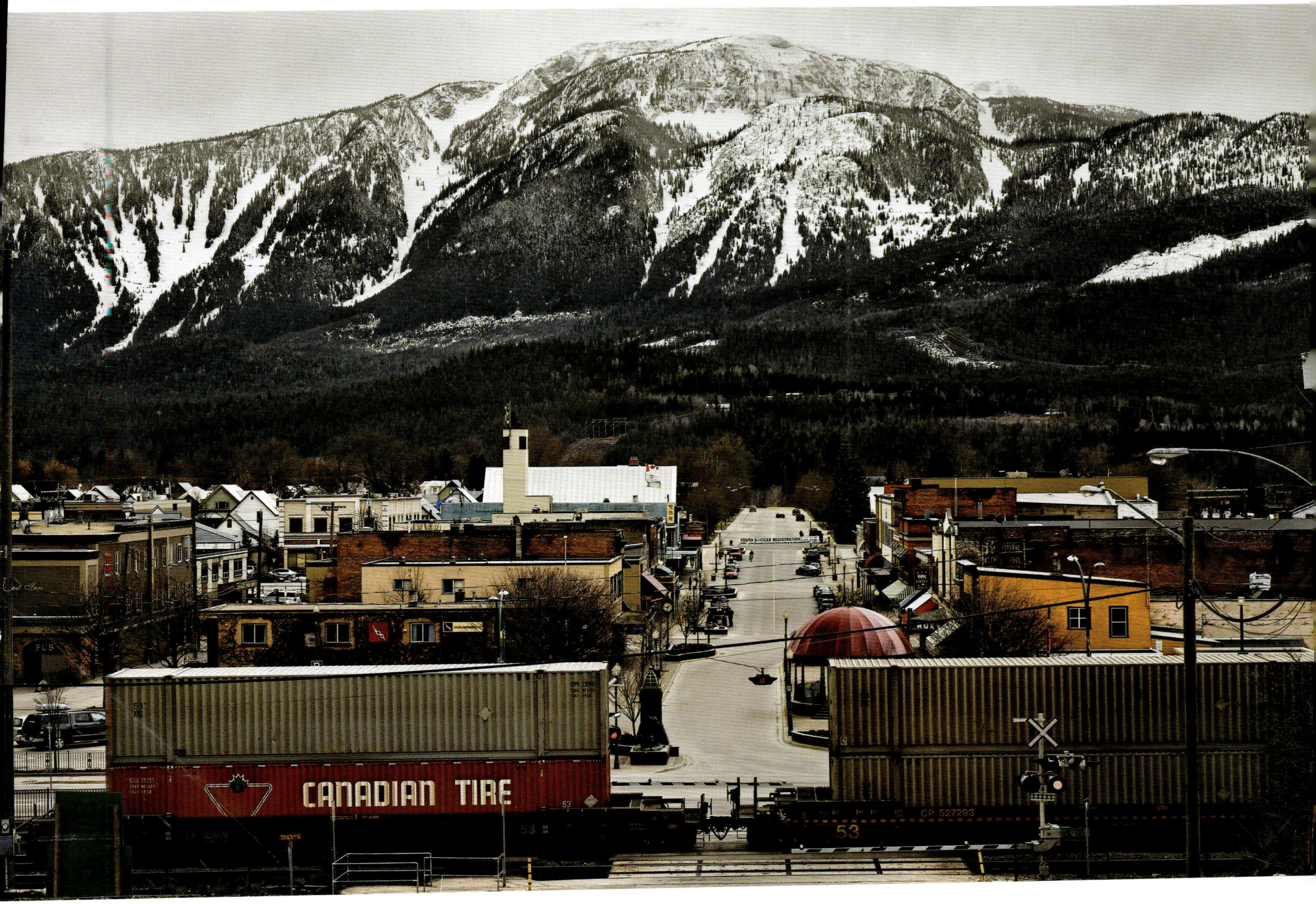

Revelstoke, Revelstoke BC

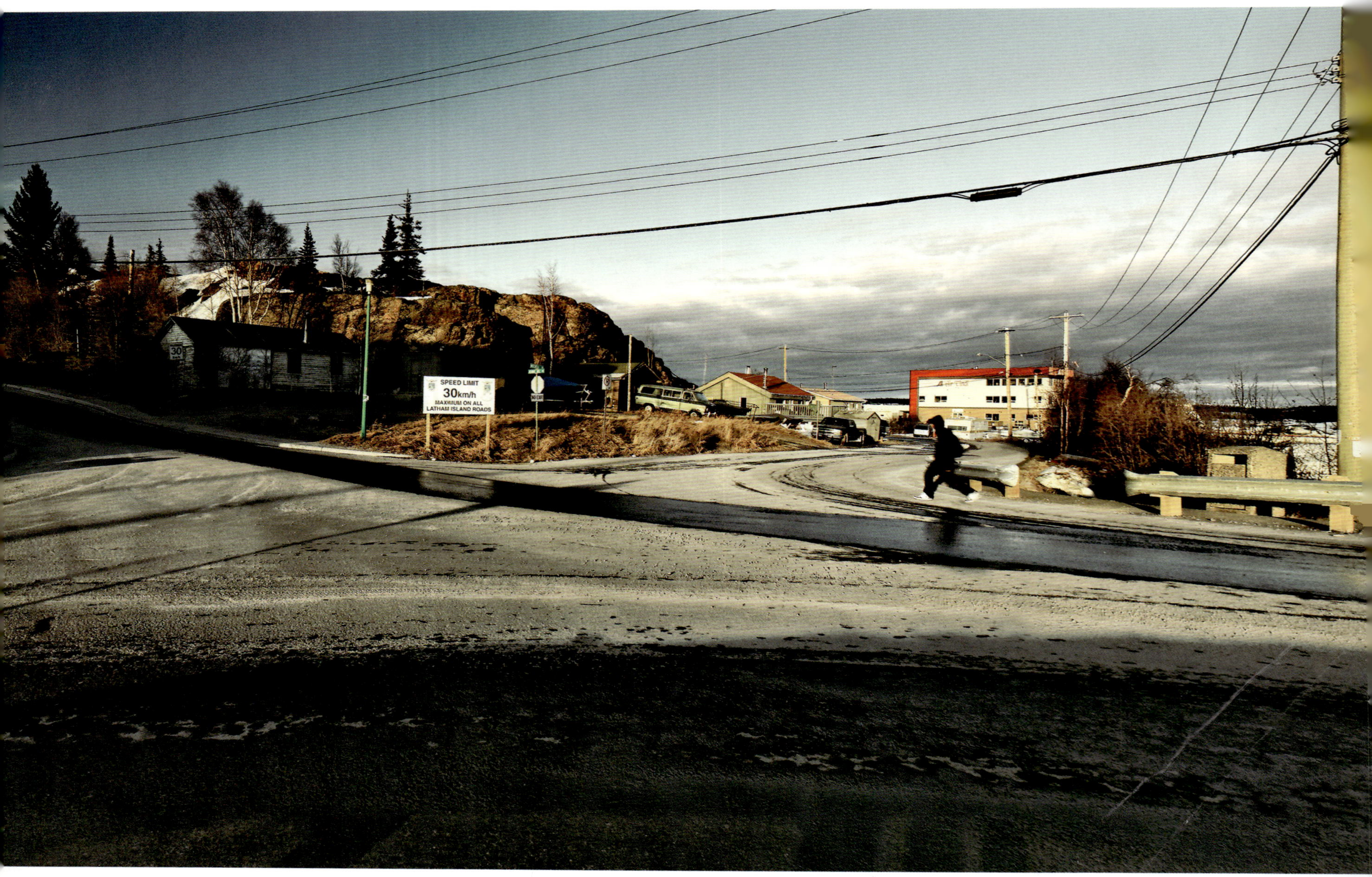

Intersection, Latham Island, Yellowknife NT

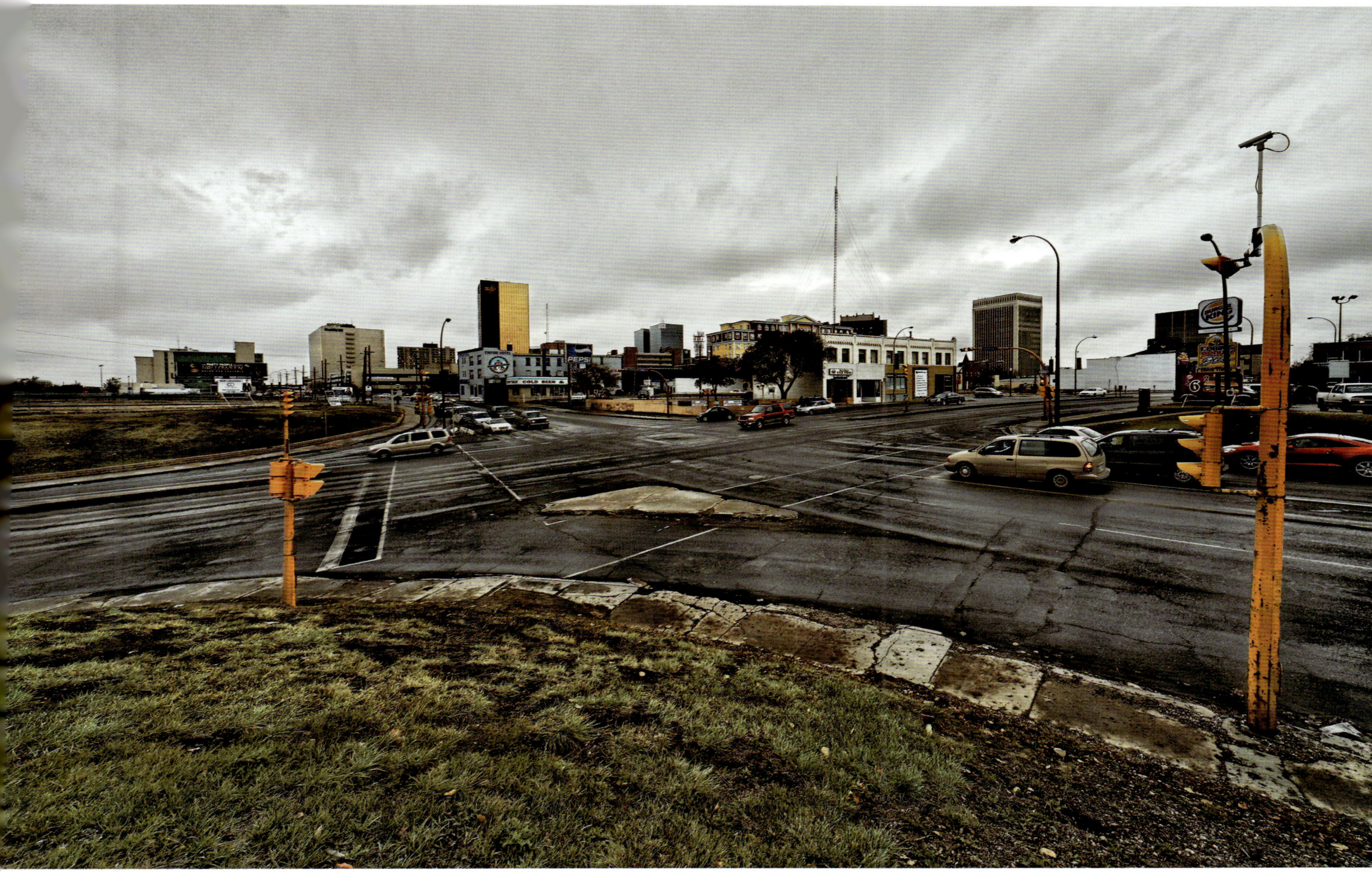

Saskatchewan Drive and Albert Street, Regina SK

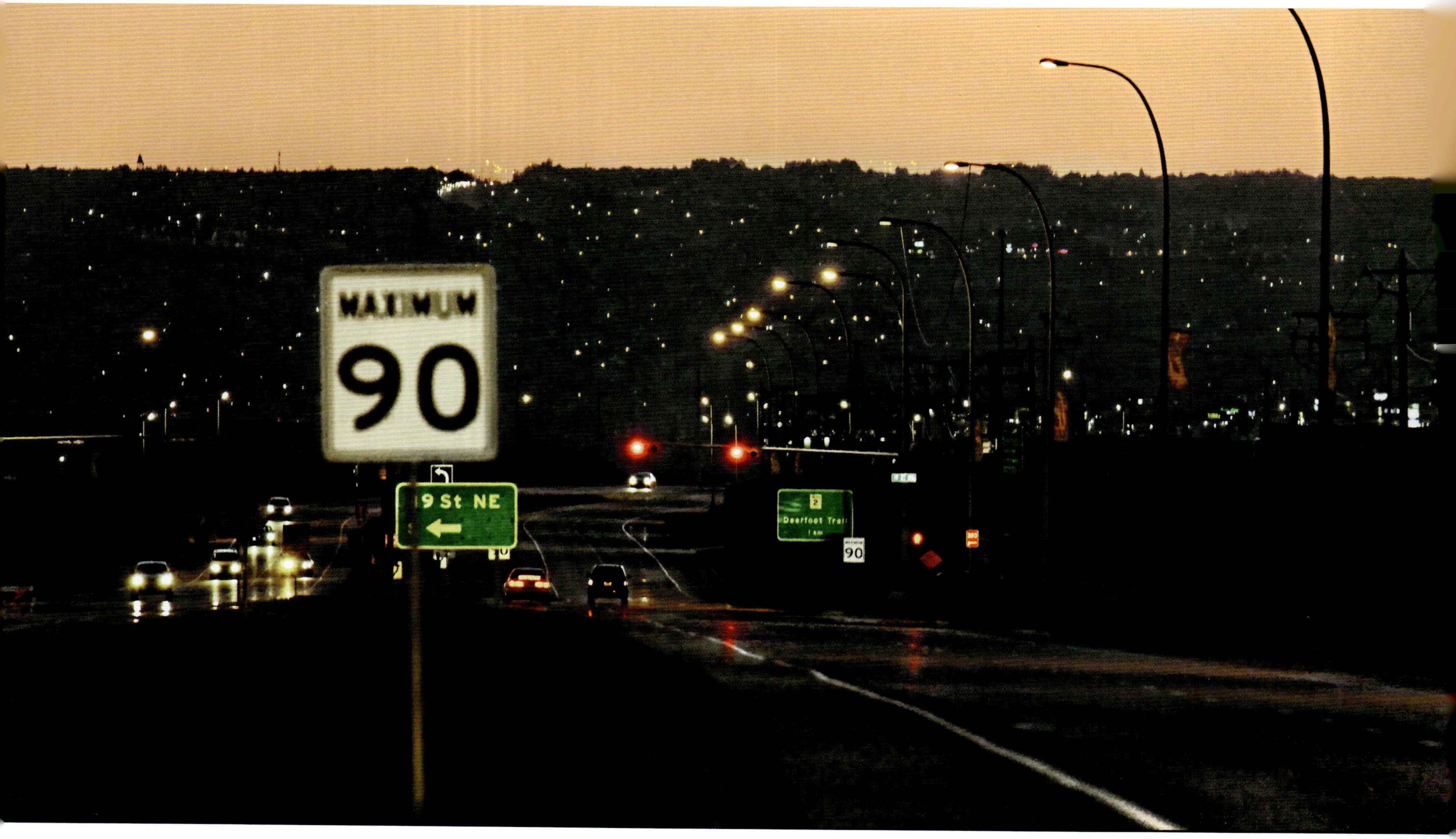

McKnight Boulevard NE, Calgary AB

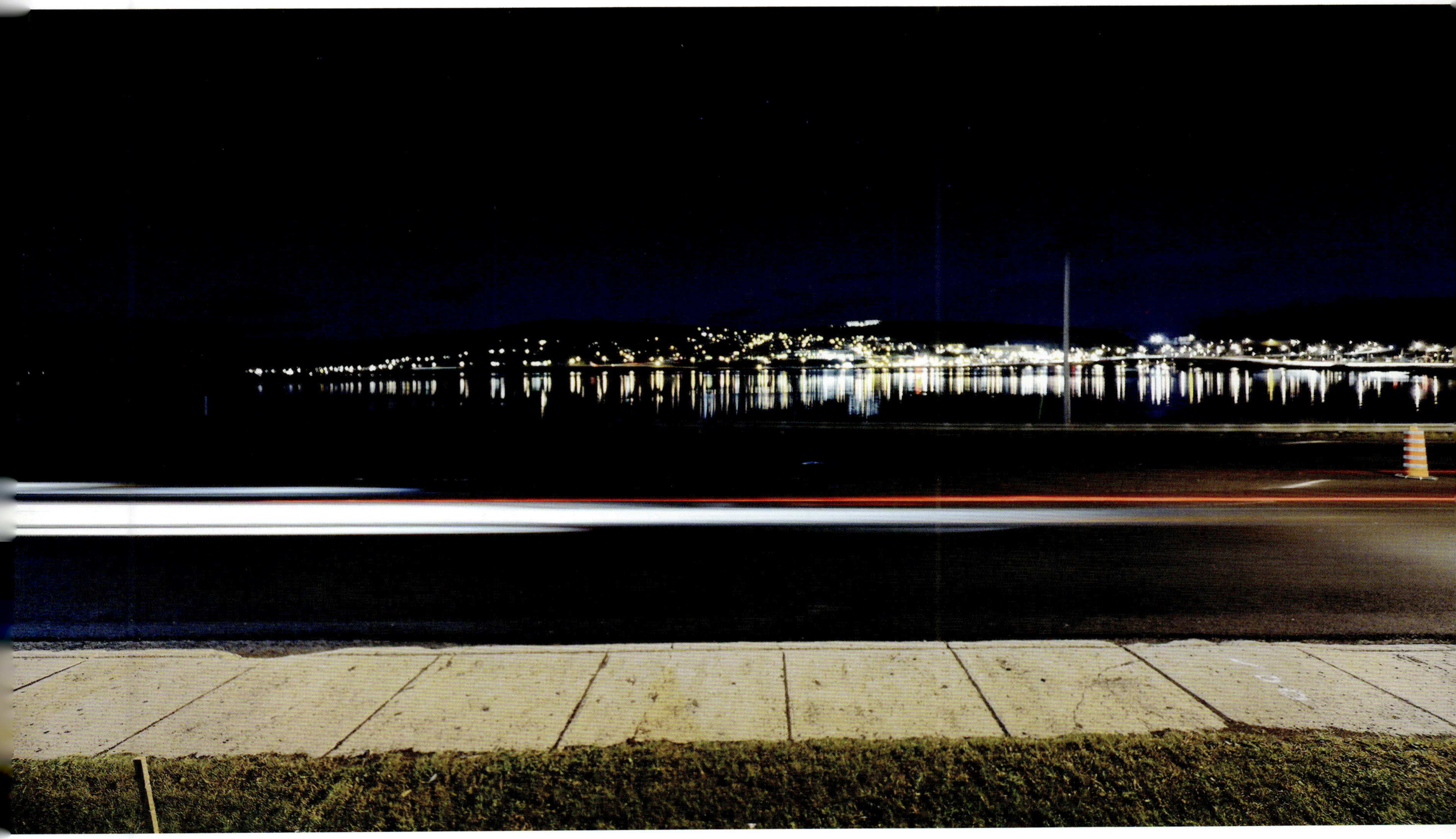

Boulevard de York Est, Gaspé QC

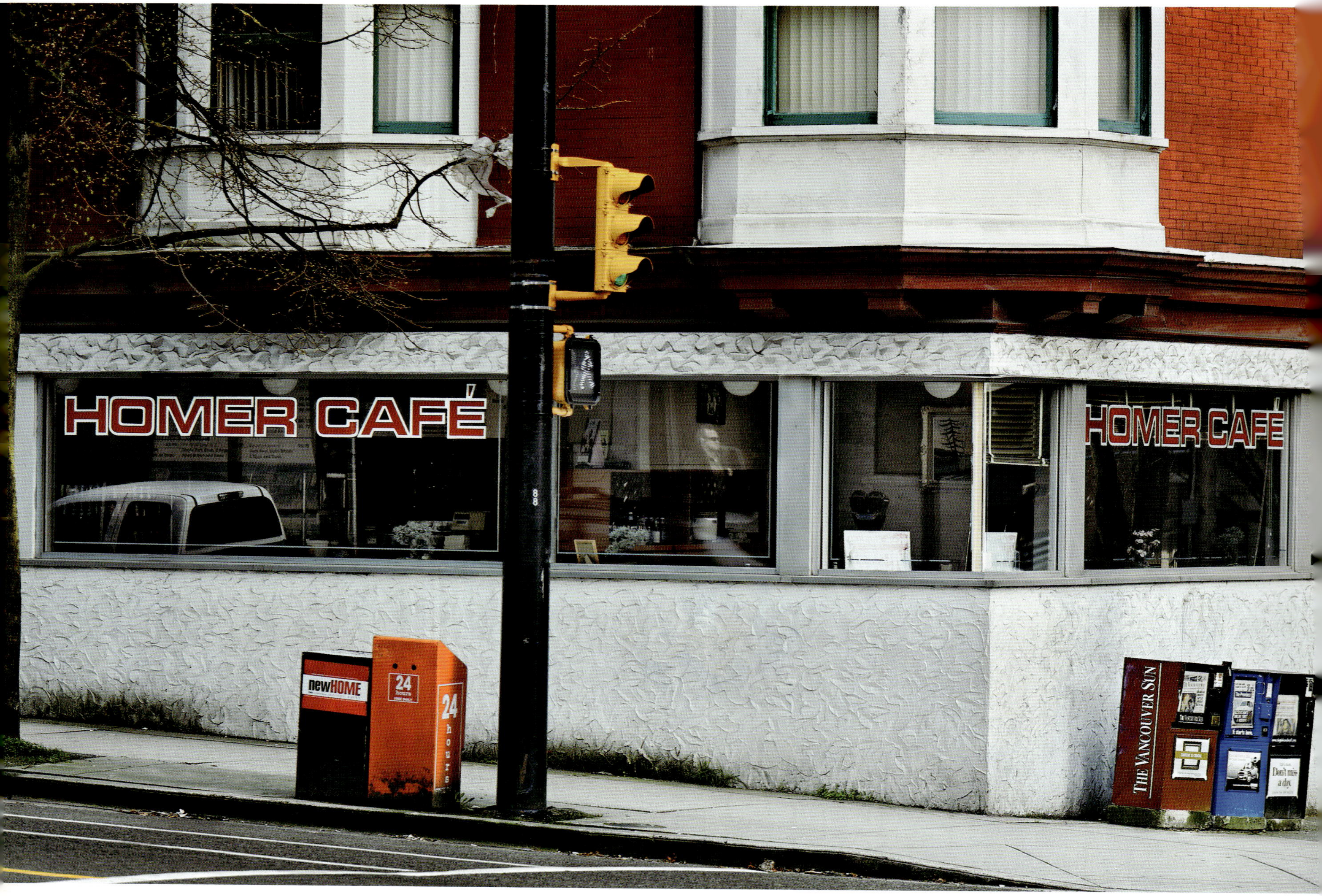

Homer Café, Vancouver BC

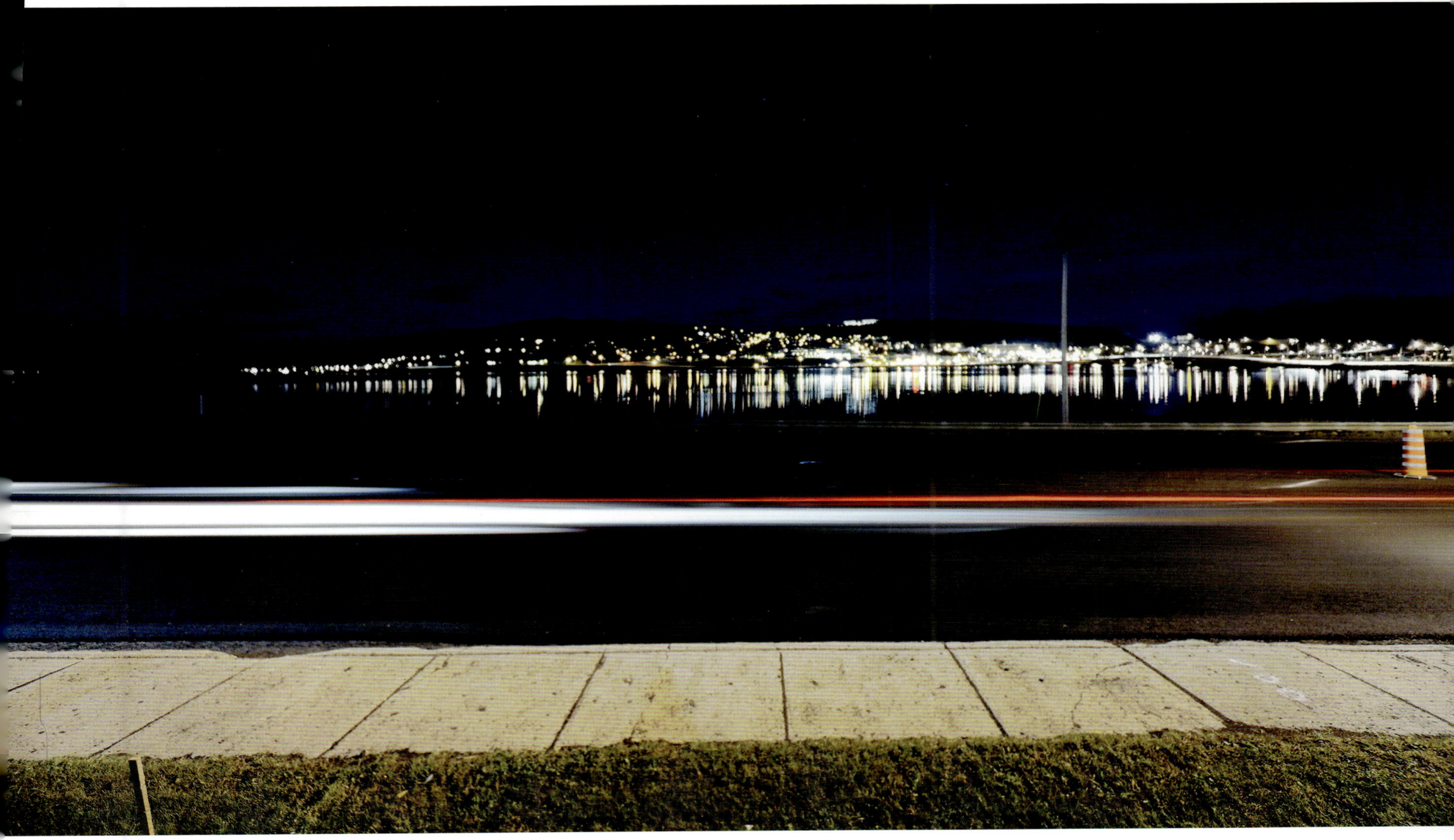

Boulevard de York Est, Gaspé QC

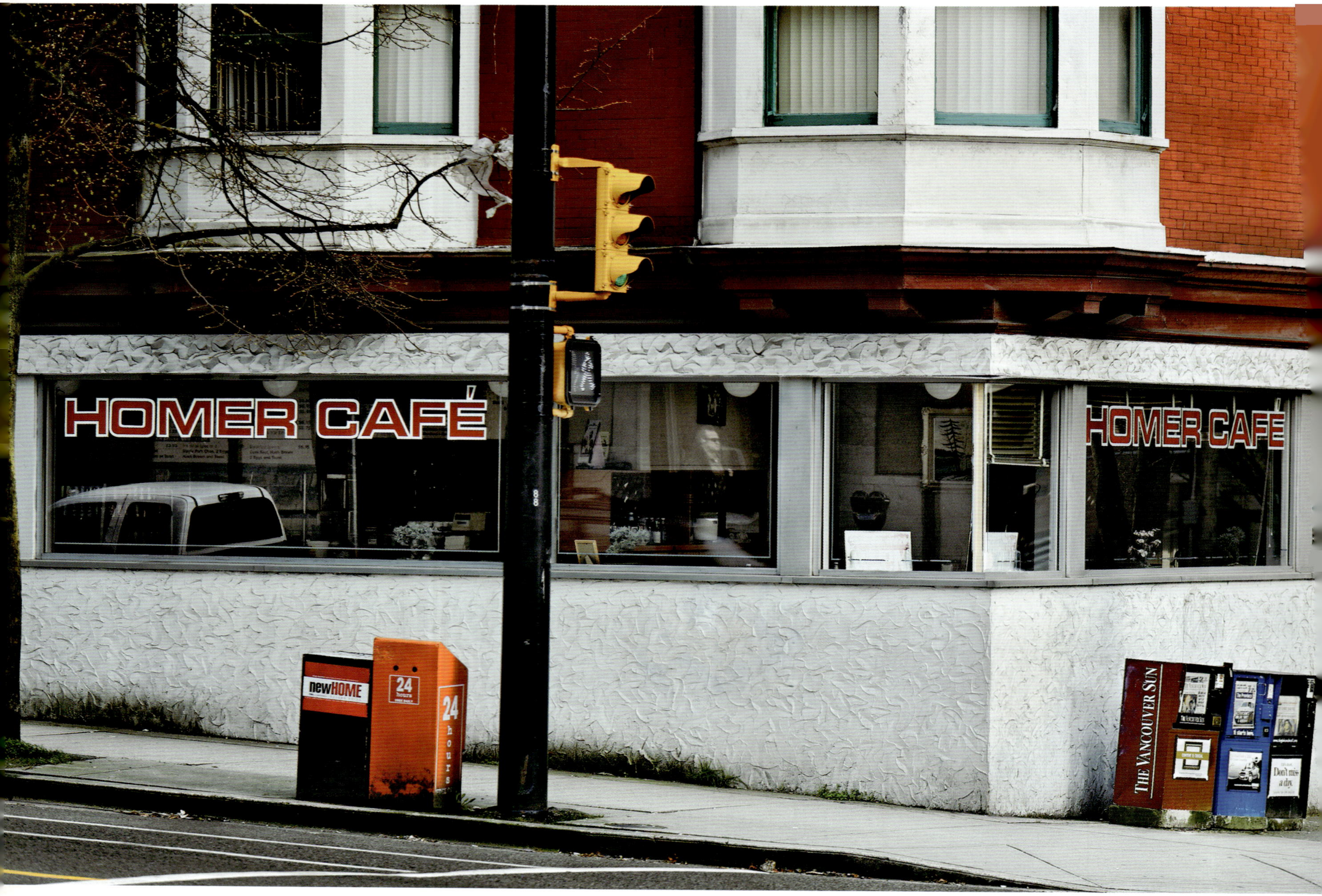

Homer Café, Vancouver BC

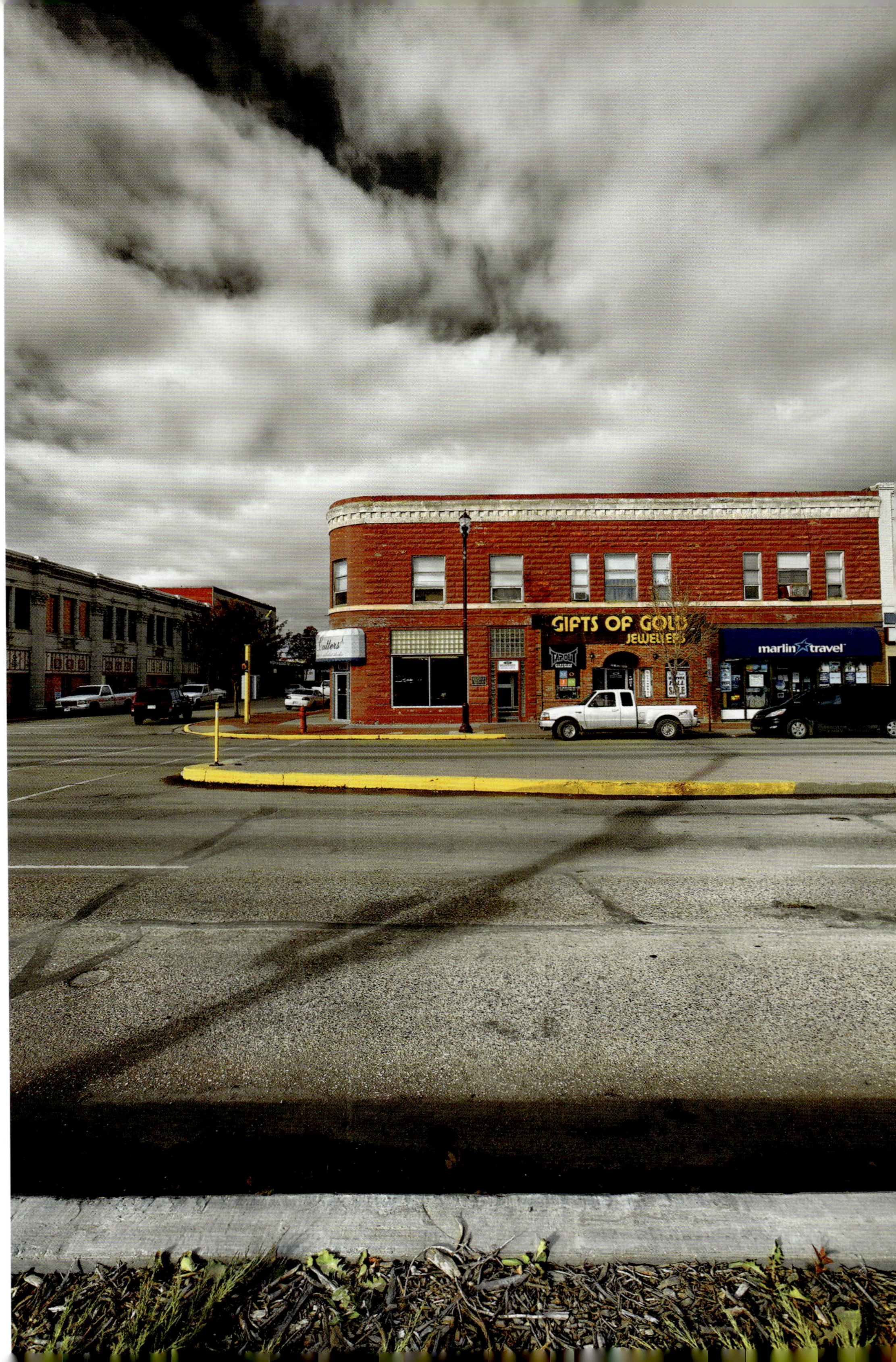

Gifts of Gold, Yorkton SK

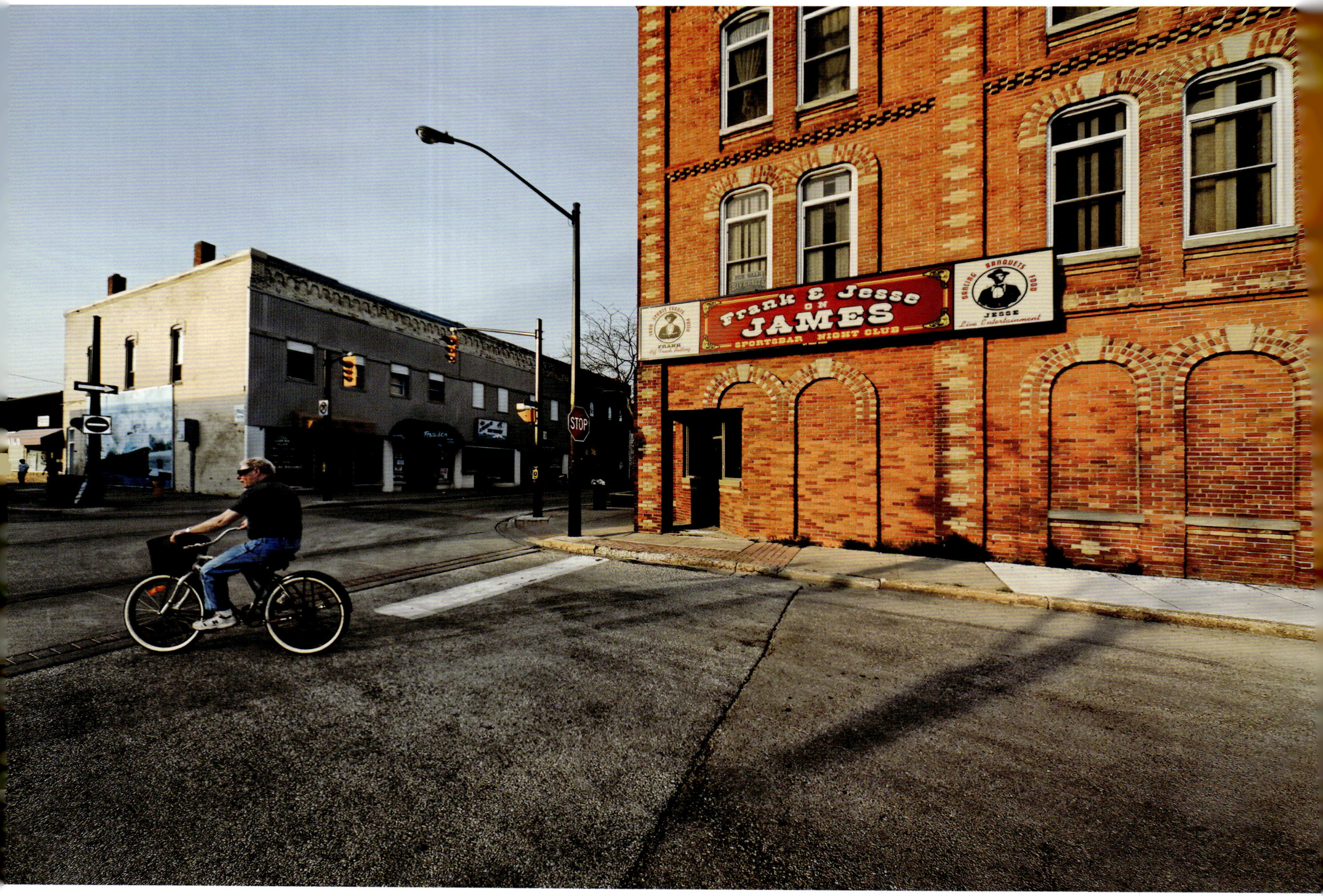

Cyclist, Wallaceburg ON

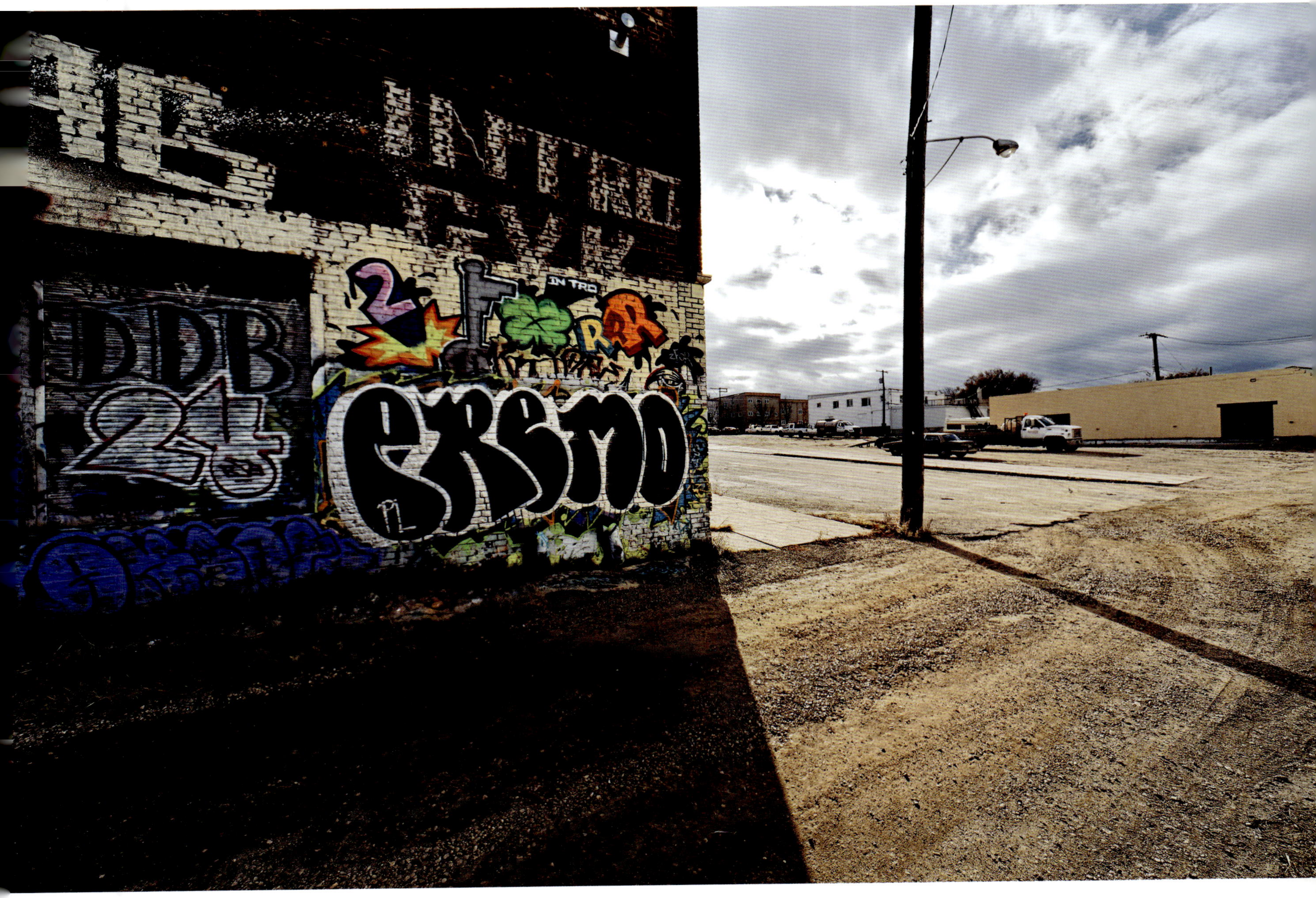

Wall, Brandon MB

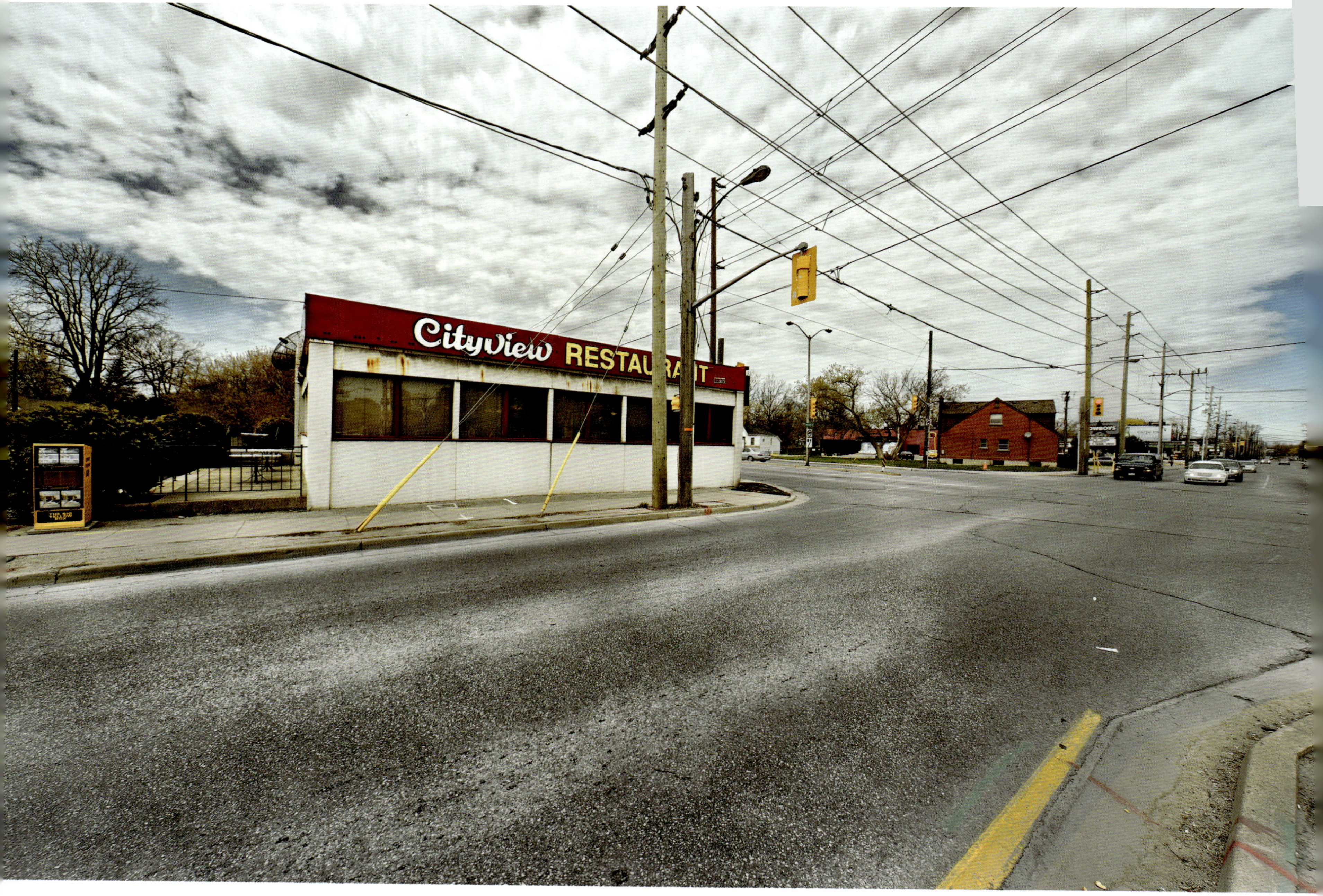

Cityview Restaurant, London ON

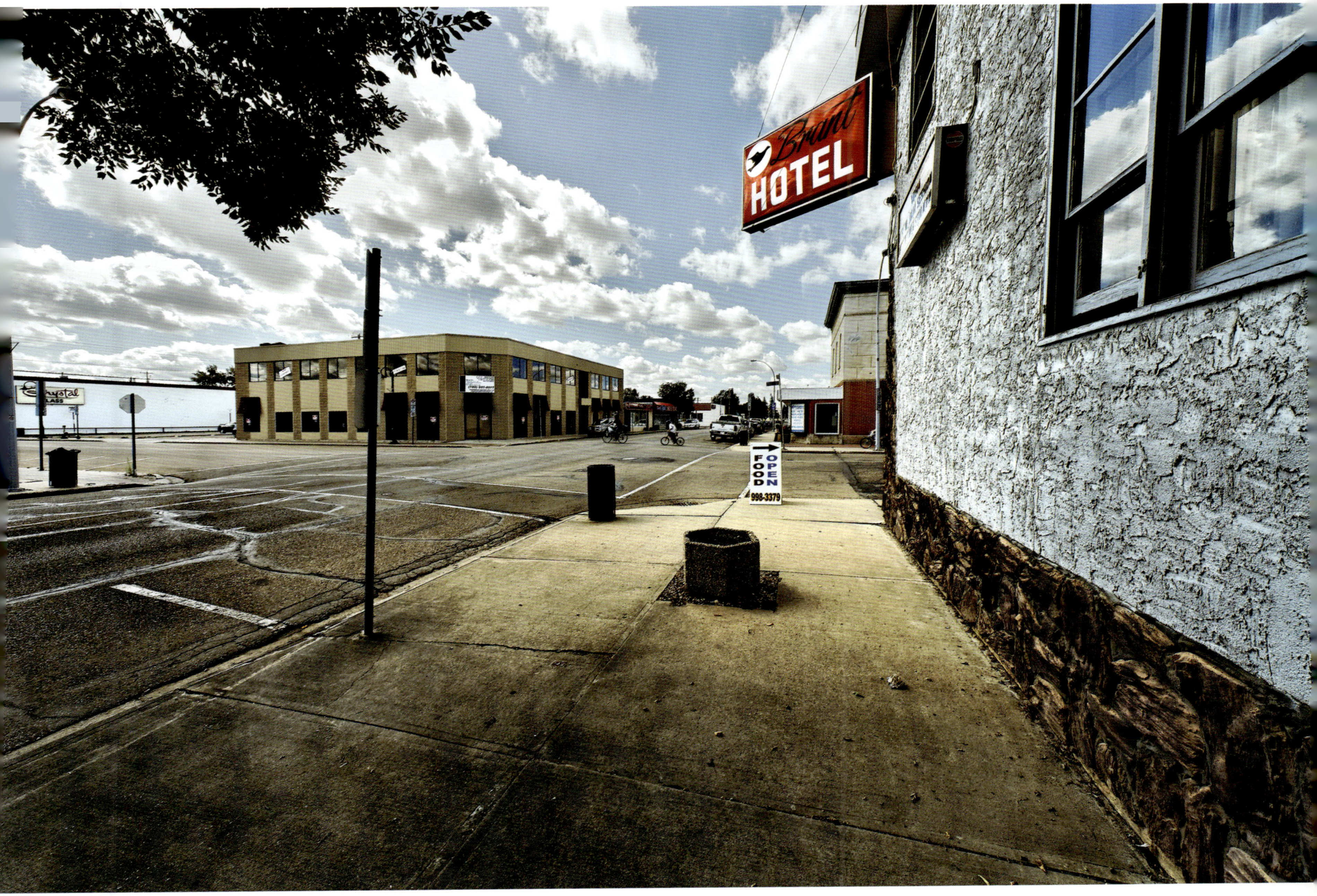

Brant Hotel, Fort Saskatchewan AB

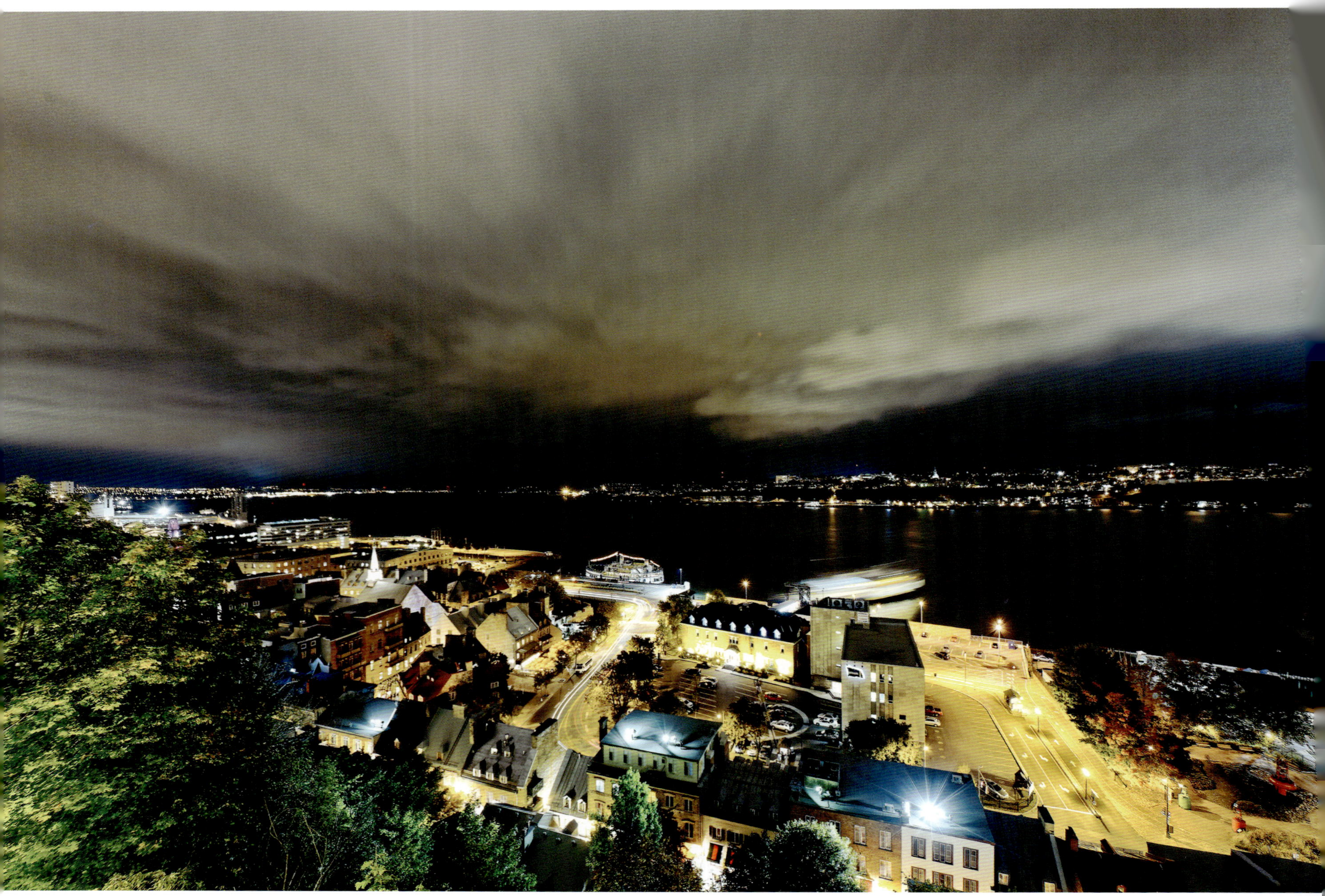

Dufferin Terrace, Quebec City QC

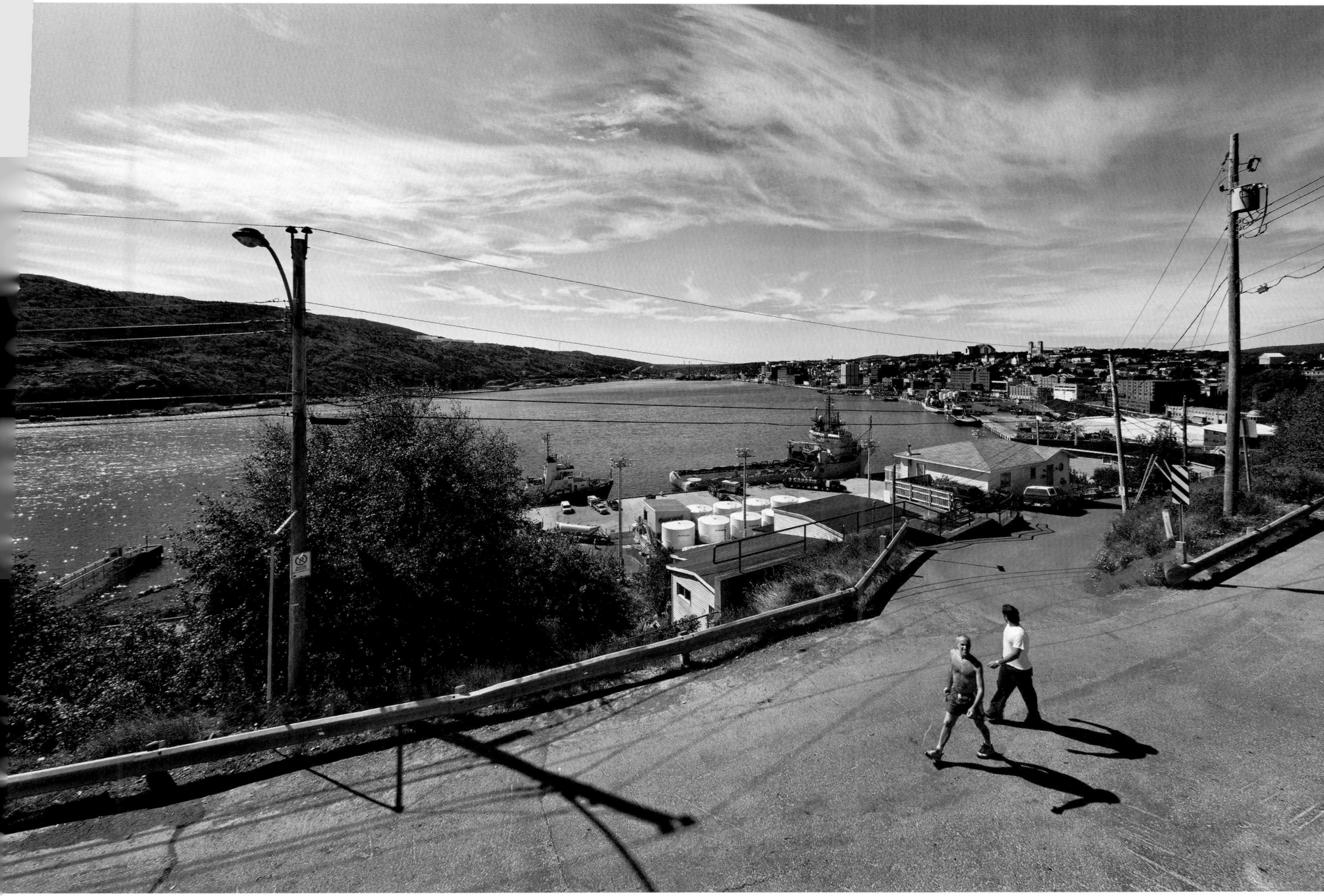

St. John's Harbour, St. John's NL

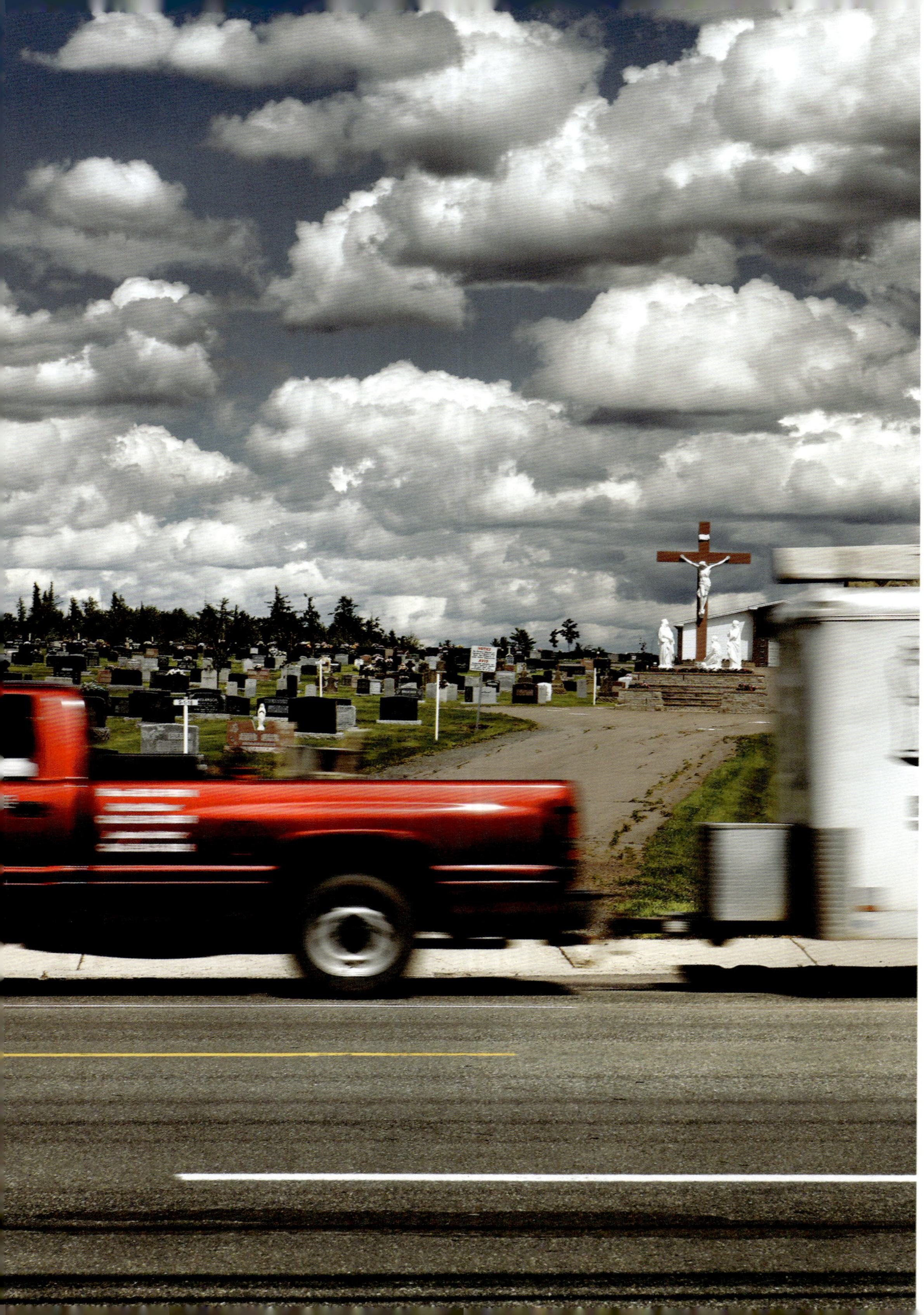

Our Lady of Calvary Cemetery, Chartersville NB

Index